Competitive Technical Intelligence

Competitive Technical Intelligence

A GUIDE TO DESIGN, ANALYSIS, AND ACTION

Mathias M. Coburn

American Chemical Society
Washington, DC

Oxford University Press
New York   Oxford

1999

Oxford University Press

Oxford   New York
Athens   Auckland   Bangkok   Bogotá   Buenos Aires   Calcutta
Cape Town   Chennai   Dar es Salaam   Delhi   Florence   Hong Kong   Istanbul
Karachi   Kuala Lumpur   Madrid   Melbourne   Mexico City   Mumbai
Nairobi   Paris   São Paulo   Singapore   Taipei   Tokyo   Toronto   Warsaw

and associated companies in
Berlin   Ibadan

 Copyright © 1999 by American Chemical Society

Developed and distributed in partnership by the
American Chemical Society and Oxford University Press

Published by Oxford University Press, Inc.
198 Madison Avenue, New York, New York 10016

Oxford is a registered trademark of Oxford University Press

Library of Congress Cataloging-in-Publication Data
Coburn, Mathias M., 1936–
Competitive technical intelligence : a guide to design, analysis,
and action / Mathias M. Coburn.
p. cm.
Includes bibliographical references and index.
ISBN 0-8412-3515-5
1. Technology—Management.   2. Research, Industrial—Management.
3. Business intelligence.   4. Strategic planning.   I. Title.
T49.5.C623   1999
658.5'14—dc21   98-51142

9 8 7 6 5 4 3 2 1

Printed in the United States of America
on acid-free paper

Preface

The principal objective of this book is to provide the reader with the ability to design, obtain necessary support for, and implement a competitive technical intelligence system in a way that enables one's organization to identify, acquire, and utilize new technology faster and better than the competition, for competitive advantage in the marketplace.

Armed with the frameworks, systems, and practices described in this book, the technology-leveraged company will be better prepared to take advantage of growth initiatives while managing risk. The company will also sharpen its ability to tap into new sources of technology in developing countries, universities, and government laboratories, as well as the technologies of well-established companies in the more technically sophisticated parts of the world.

Primary beneficiaries of this book are R&D management and those practitioners within R&D charged with the responsibility of carrying out technology intelligence programs, strategic technology planning, and acquisition of external technology. Secondary audiences for the book are those responsible for running a technology-leveraged business, as well as their marketing, planning, and information system functions, as well as anyone else in the organization who participates in the strategic management of technology or is concerned with its impact on the business.

Corporate America is now in the transition from seeking technical self-sufficiency to finding ways to leverage their existing resources with external technology in order to grow their business. Unfortunately, corporate America has been relatively slow, compared to corporate Europe or Japan, to learn that, given increased global competition and speed to commercialization, they can no longer single-handedly support the increasing cost and complexity of generating significant new technical advances, as well as the broader technical base required to support an expanded line of products or markets. Although there are good reasons for corporate America's lagging position it doesn't change their need to catch up with the rest of the world.

For those concerned with science and technology policy in government or academia, the methodology presented for identifying and assessing new technologies worldwide is equally relevant. The book also addresses other relevant topics including the establishment of intelligence networks, frameworks for technology transfer and strategies for carrying them out, as well as maximizing benefits from total global resources.

The book provides the reader with abilities based upon new principles backed up by unique, proven experience not readily available elsewhere. It can be used as a text to help provide the skills and knowledge needed by new generations of technology managers.

Readers will note the very limited number of literature citations in the book. Currently, there is little literature available that discusses either techniques or a systematic approach to competitive technical intelligence. Much of the material in this book was developed by me, based on my personal experiences which were devoted to the acquisition of external technology, during a 13-year period of my career at DuPont. At that time, the businesses I supported were going through several transitions; from an internal to external focus, from a technology follower to technology parity (if not a leader), and from multinational to integrated global businesses.

Initially, the approach to identifying new technologies of interest through competitive technical intelligence was empirical, sometimes even accidental! Nevertheless, some early positive results were obtained that proved to be very influential for the businesses. This process also involved a lot of trial and error, false starts, and dead ends. Fortunately, none of these approaches were fatal. With increasing success, and contributions to the effectiveness of the overall R&D effort, came additional funding which was used to refine the frameworks, processes, and techniques, as well as to benchmark other companies. These benchmarking activities added perspectives to the program while reinforcing much of what was being done.

Later, the empirically developed material was integrated with concepts and frameworks being utilized by The Fusfeld Group, a technology management consulting firm, which further validated the methodologies previously developed, provided additional insights into the rationale for why things work the way they do, and positioned the information obtained to become even more actionable. As a consultant with The Fusfeld Group, I quickly discovered that the methodology developed was generic, and not limited to the situation at DuPont, nor to the chemical, petroleum, and allied industries. The frameworks and processes have been applied with equal success in such diverse industries as automobile components, mining and metallurgy, electronics, aerospace, public utilities, as well as consumer and personal-care products.

I regret that I was unable to use the actual names of many of the companies and technologies uncovered by the methodologies described. Hopefully, this will not detract from the utility of the material. The examples cited are real, but competitive technical intelligence is something still not openly discussed by many organizations. On several occasions, after conducting a public workshop, I would be surrounded by participants eager to discuss potential applications in their respective organizations, *but not in front of the others present.*

Today, there is a growing trend towards increased comfort levels among organizations to openly discuss their activities regarding competitive technical intelligence. Perhaps case studies in future publications will be able to cite actual names of companies and technologies. Meanwhile, as you go through the examples in this book, I ask you to fill in your own ideas regarding relevant company names and technologies.

Acknowledgments.   First, I thank Dr. Herbert I. Fusfeld, chairman of The Fusfeld Group, and the author of a number of publications including the first in this series titled, *Industry's Future: Changing Patterns of Industrial Research.* It was Herb Fusfeld's idea to create this book, and it was his belief in my knowledge and abilities that provided the necessary encouragement for me to accept this challenge.

A thank you is also in order to Alan R. Fusfeld, president and CEO of The Fusfeld Group. Alan has been my coach and counselor for a long time, starting with my later years at DuPont. It was Alan who helped me to understand the generic nature of the frameworks and methodologies, and who was also responsible for showing me how competitive technical intelligence is linked to strategic planning and development.

Finally, I wish to acknowledge the late Dr. Arthur W. Andresen, formerly an assistant director of research at DuPont who provided me with direction and encouragement to develop a strategic patenting process which subsequently led to the development of the competitive technical intelligence methodology described in this book.

Contents

Competitive Technical Intelligence

The Management of External Technology

The Increasing Role of External Technology

In his book, *Industry's Future: Changing Patterns of Industrial Research*, Dr. Herbert Fusfeld points out that, "from 1900 into the 1970s, the growth of industrial research can be seen in retrospect as a steady increase in the technical self-sufficiency of corporations".[1] He further states, "sometime in the mid-1970s progress toward technical self-sufficiency began to reverse . . . Corporate requirements for technology steadily outgrew the capacity of internal technical resources."[1] From the 1970s on, corporations regardless of size could no longer afford to be self-sufficient with regard to technology, and needed to augment their internal efforts with technology acquired from outside.

There are several reasons for the need to augment internal industrial research. The significant increase in globalization of industries has not only brought more competitors into an expanded marketplace but also has intensified competition from not-in-kind and substitute products. With this increase in competition, product life cycles have decreased significantly while cycle time from product concept to commercialization has also undergone substantial time compression. Corporations around the world do not have the luxury of conducting R&D at their own pace. Technology acquisition saves time!

Additionally, Fusfeld points out that to maintain a competitive position, corporations find that a broader technical base is required to support an expanded line of products or markets, particularly as companies become global competitors. They also have to bear the increasing cost and complexity of generating significant new technical advances. This cost has become increasingly unbearable not only in terms of dollars, but also in terms of the number of people needed to carry out the research and development.

In the 1990s, the issue of leveraging internal efforts has become exacerbated as a result of downsizing or reengineering the corporation. Today, executives are faced with the dilemma of meeting corporate goals to grow their businesses at rates that cannot be accomplished from their existing businesses alone. At the same time, their internal resources have been reduced. This limitation of resources hampers their ability to develop new products and technologies to grow these existing businesses, let alone generate new businesses. To meet this challenge with reduced resources, it is imperative to leverage efforts with external technology!

Barriers to Utilization of External Technology

Although the barriers are coming down, obstacles still exist within many organizations to utilizing external technology to leverage internal efforts.

First, there is the issue of *will*. After decades of focusing on technology self-sufficiency, it is difficult to quickly make the transition to interdependency. The will or desire to make this shift must come from at least the chief technology officer if not from the chief executive officer (CEO). Only a few years ago, a corporate vice-president proclaimed, "If we have to license-in technology in an area in which we are working, I will consider it a failure of R&D." It was no wonder his organization suffered heavily from the Not Invented Here (NIH) syndrome, and was exceedingly slow to accept external technology. As a result the organization suffered a loss of competitive position that nearly proved fatal, before a new administration made a complete reversal in policy.

The issue of will requires not only the desire to become externally focused, but also the will to organize, staff, and fund accordingly. Too often, key executives will espouse external focus yet turn down their subordinates' request for personnel and budgets to implement such programs. This may stem, in part, from the erroneous belief that technical intelligence and acquisition systems are expensive. They are not! This book will clarify the issue of staffing as well.

The second issue is *skill and knowledge*. How do you identify new technologies of potential interest to you before your competitors do, and before the technologies become commercial? Assuming your competitors have the same information on new external technology that you do, how do you capitalize on this technology before they do, and further convert this into sustainable competitive advantage?

External Technology as Part of a Strategic Technology Plan

Jay W. Schultz, Director of Technology for ALCOA's Worldwide Automotive Business, defines technology as "A body of knowledge, tools, and techniques, derived from science and practical experience, that are used in the development, design, production and application of products, processes, systems and services to meet customer and business needs."[2]

The essence of managing external technology is the process of selectively taking information from the external environment, converting it into intelligence, and taking action on the knowledge developed, to acquire technology in the most cost-effective manner, so that internal resources are augmented and leveraged for competitive advantage in the marketplace. Although external technology is often acquired on a reactive, ad hoc basis, it becomes a considerably more potent weapon when it is considered as a component of a complete strategic technology plan. Proactive planning sharpens the criteria for acquiring technology by defining the role of external technology in concert with internally driven programs.

Two illustrations can demonstrate the relationship of external technology as a component of a strategic technology plan. Figure 1-1 is a plot of the business impact of selected technologies versus the organization's competitive position in each of them. A high business impact results in increased

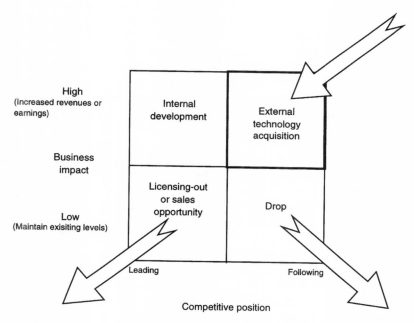

Figure 1-1. Influence of business impact and competitive position on external technology considerations.

revenues or increased earnings, while a low business impact results in no change in sales or earnings.

This chart indicates that technologies that have the potential for high business impact, and in which the organization has a leading competitive position, are the best candidates to be driven by internal development. These technologies offer the opportunity for differentiation and competitive advantage.

Alternatively, technologies in which the organization enjoys a leading competitive position, but which have a relatively low business impact, may provide greater impact and relevance in other industries or businesses outside the scope of the enterprise. In such cases, these technologies represent valuable intangible assets for potential sale or licensing out.

As for technologies in which the enterprise is a follower, if the potential impact is likely to be low, the program is a candidate to be dropped in favor of higher impact programs.

Finally, technologies in a following position, that have the potential for high business impact, are candidates for acquisition or external leveraging. These technologies should become the focus of external technology management! Even if the enterprise has resources available to develop these technologies internally, it would usually be wasteful to deploy resources in this manner. Not only is it unlikely that the enterprise can develop a differentiating position, but a greater-than-normal effort is likely to be needed simply to catch up to competition. Even if this position is achieved, it may not be accomplished in the time frame necessary to capitalize on the technology.

One of the myths regarding external technology is that it does not require utilization of internal resources. Acquisition of external technology saves precious time, but internal resources are always required to bring the technology into the enterprise. Furthermore, these resources need to be knowledgeable in the external technology if the technology transfer is to be effective. The amount of internal resources will depend on the maturity of the technology. For example, much less internal R&D will be needed for a mature turnkey operation compared to an emerging concept.

Another tool to identify external technology opportunities is shown in Figure 1-2. In this matrix, risk is plotted against business impact. Risk is defined as the combination of both technical and commercial risk. Technical Risk considers both the probability of technical success and the associated research and development costs. Commercial risk addresses the ease or difficulty in gaining customer acceptance of the new product, process, or service.

High-impact/low-risk projects clearly should be high-priority internal programs. These programs offer the best opportunities for differentiation from competition, competitive advantage, and potential increase in revenues or earnings at relatively low commercial and technical risk. Unfortunately, the more usual case is that high-impact programs are also high-risk, from a technical or commercial standpoint or both. This case tends to be particularly true in mature industries. In these instances, the probability of success may

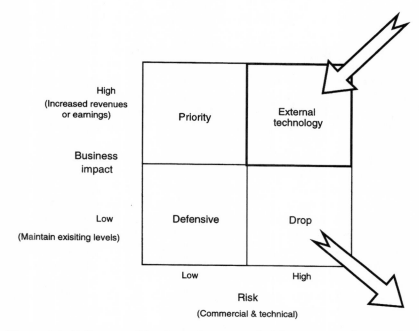

Figure 1-2. Influence of business impact and risk on external technology considerations.

be increased through a strategic alliance with a partner that has better market presence or technical competency in the specific technology of interest. In addition, if a strategic partner shares the development cost, the financial risk is reduced.

As demonstrated with these matrices, external technology opportunities can be best identified within the context of a strategic technology plan. This plan enables an organization to move from reactive to proactive behavior in terms of acquiring external technology and/or seeking external strategic alliances. The technology plan becomes the driver for action. Identifying appropriate sources of specific technologies and appropriate partners to develop specific technologies requires an effective competitive technical intelligence function. If the strategic technology plan is the driver, competitive technical intelligence is the tool!

External Technology and Competitive Technical Intelligence

If intelligence regarding external technology is to be actionable, it is either incorporated into a tactical or strategic planning process to improve the organization's own competitive position, or it becomes the basis for acquisition of new technology.

Surprisingly, profiling a competitor's technology deployment is an excellent way to identify new technology! This discovery was initially made empirically. In the course of profiling a leading competitor, a new emerging technology of considerable interest was identified. In subsequent searches, it was further discovered that others were also working in this new area, and ultimately this new technology platform was acquired from a noncompeting third party. In retrospect, perhaps competitor profiling as a source of new technology should not have been so surprising. It is reasonable to find competitors engaging in researching new technology in areas of their core competencies.

There are several implications of this finding. First, it enables the investigator to accomplish two objectives with the same search protocol; profiling key competitors for strategic or tactical planning purposes, and developing leads to potentially new technologies. Second, competitor profiling allows for a cleaner, quicker, more efficient search protocol, minimizing the amount of irrelevant data collected. While any search should not necessarily be limited to competitor profiling, consideration should be given to including it in any search protocol.

An example of competitor profiling occurred with a domestic consumer products company that was concerned that its global competitors were developing new technology (and subsequently new products) in an area considered to be a core business for this domestic company. The company believed that its multinational competitors may be developing this technology in their overseas laboratories using technology originating in Europe or Asia, to which the domestic company did not have direct access.

Subsequent competitor profilings of these global competitors showed only a minor effort being expended in the area of interest. However, a subsequent search for this new technology identified a major domestic supplier to the consumer products industry, as being the principal developer. The conclusion became clear. The domestic company could maintain competitive parity with its larger rivals by simply working closely with this domestic supplier. There was even the possibility of developing some lead time if a joint development effort could be worked out with the supplier, thereby preempting competition.

Sources of Technical Intelligence versus Sources of Technology

The reader may sense that the activities discussed relate only to industrial companies. This is not so. Universities, institutes, and government laboratories also need an awareness of what is going on in technologies of interest outside of their own organizations, although the reasons may differ somewhat from the industrial firm. For example, having an external awareness could enable an institution to upgrade its own direction for future technical work. In addition, company profiling could help identify potential clients for technology under development in academia, an institute, or the government sector.

In subsequent chapters, sources of competitive technical intelligence will be described. These represent sources of *information*, not sources of *technology*. The information sources will, in turn, lead to identifying the sources of the technology, which could be industrial companies, universities, institutes, or government facilities. These organizations, just as private companies, apply for patents, publish articles, and send their people to conferences and seminars. As a result, the search methods described in this book will find the sources of technology regardless of which sector of the economy they fit into and regardless of geographical location.

Here's an example. A new catalyst technology of interest to a North American chemical company was identified in one of the scientific journals, as well as in several pre-patent publications, as being developed by an Australian research institute. Subsequently, a joint development program was established between the organizations which would enable both to participate in the commercialization of ensuing products, with the institute also receiving funding from the chemical company to hire additional people to expand the field of research. The chemical company was therefore able to leverage its own internal resources with those of the institute.

GUIDING PRINCIPLES

Principle #1: Today, every enterprise needs to avail itself of external technology to leverage internal efforts.

Principle #2: A cost-effective way to identify new technologies is to first profile your competitors' technology.

2

An Overview of Competitive Technical Intelligence

What Is Competitive Technical Intelligence?

The definitions of competitive intelligence and competitive *technical* intelligence are basically the same. With only slight modification, a definition for competitive technical intelligence (CTI) can be derived from one developed by Kirk Tyson, Chair of Kirk Tyson International. CTI is defined as "The analytical process that transforms disaggregated competitor technology data into relevant and usable strategic technology knowledge about competitors' positions, size of efforts, and trends." The information can relate to past, current, or future time frames. While the focus will be primarily on the future, historical information is useful to plot and analyze trends.

The Three Elements of the CTI Activity

The three basic elements of the competitive technical intelligence activity are *data collection, data analysis*, and *action*. The components of these elements are shown in Table 2-1.

Data Collection. A popular misconception among people new to data collection is that relevant data is scarce and hard to come by. To the contrary, the problem is that there is so much data out in the public domain, that unless one focuses the search, one can figuratively drown in data! Larry Kahaner, author of the book *Competitive Intelligence*, said, "Information is a com-

Table 2-1 Elements of Competitive Technical Intelligence

Data Collection	Data Analysis	Action
Design is critical	Design is critical	Belief in the data
Actionable level of accuracy	Judgment	Trust in the analysis
The foundation of a complete story	Insight	A natural fallout from the analysis, if the design was adequate
	A consistent story	
	Putting the competitive puzzle together	

modity—the greater the supply, the lower the value. The real worth of information comes when it is turned into intelligence."

The idea, therefore, is to collect no more data than will be needed to take action. This idea assumes that the purpose for the study has been clearly defined with an up-front vision of possible actions resulting from the work. Design is critical, and design starts with a good understanding of the purpose of the study. Without this understanding, there is a tendency to collect more information than is probably needed, let alone the possibility of not collecting all the relevant information at that.

Data Analysis. The design is also critical to the analysis phase of the work. As the data is collected it can be plotted on trend graphs, tabulated in charts, or placed in analytical grids, which have been anticipated as part of the design. This facilitates analysis, and permits insights to be readily developed. Utilizing a predesigned analytical framework also facilitates the development of a consistent story, which strengthens credibility in the eyes of the decision makers. CTI analysis involves putting together a competitive jigsaw puzzle of facts. Analytical frameworks foster integration of information.

Often, it is easy to draw conclusions from graphs, charts, and matrices. Nevertheless, it is incumbent upon the analyst to provide insights into what the data is indicating, and why the competition may be behaving in the manner described by the analysis.

All the analysis in the world cannot be a substitute for good judgment. Just because a competitor is taking a certain course of action, does not mean that the same action is appropriate for another organization. It seems to be human nature to assume that the competition is always right, when there is a difference in approach between one's organization and the competition. A more complete perspective can be obtained by utilizing more than one source of CTI, to aid the decision maker in ultimately deciding to stay the course or to react to the competitor's initiative. Once the analysis is completed, judgment will still need to be applied in the decision-making process, in order to take subsequent action.

Action. No action will be taken based on CTI analysis unless the three conditions for action cited in Table 2-1 are met. First, the decision makers must have a firm belief in the data. They must accept the sources of the information as being valid. Questionable sources will ultimately lead to uncertainty regarding the analysis and conclusions, resulting in reluctance to make decisions, let alone to take action.

Second, decision makers need to trust the analysis. The information needs to be presented in a logical, understandable fashion, so that the decision makers understand the assumptions, and are able to follow the flow of the analysis to its logical conclusion. Equally important, the decision makers must trust the people carrying out the data collection and analysis. Staffing of the competitive intelligence function needs to be done with this consideration in mind. If the data collectors and analysts are not trusted, the wrong people have been assigned to the function.

Finally, the results and conclusions need to be a logical fallout from the analysis. This comes back to the design of the study. When designed properly, the analytical frameworks are clear, concise, and easy to understand. Conclusions become a natural fallout from the analysis and are compelling. The more compelling the fallout, the more likely that the conclusions will be acted upon.

The true test of the decision makers' belief in the data, trust in the analysis, trust in the analysts, and inclination to take action, occurs when the intelligence developed differs significantly from the decision makers' preconceived ideas of the competitive outlook. Unless the CTI activity can obtain this level of acceptance from the decision makers, the effort is not worth undertaking, as it will remain as an exercise in data collection and analysis.

A Case of a Lack of Credibility

The senior management of a worldwide supplier of materials to automotive original equipment manufacturing (OEM) assembly plants believed their company to be number two out of the three global competitors and a close follower to number one, with regard to introducing new technologies. An internal competitive technical analysis clearly indicated that the company was not a close follower at all, but rather a distant third out of three, with the gap widening each year. This finding was supported by information showing that the two leaders were consistently outspending this company with regard to R&D. If this trend continued, it would only be a matter of time before this company would be out of business!

Even after recovering from this shocking news, the senior management refused to accept this study, because it differed too radically from their own opinions. However, they also realized that they could not simply ignore it. As a result, an outside consulting firm was engaged to confirm or refute the results of the internal study. The results were subsequently confirmed, thereby encouraging the senior management to consider actions necessary to close the technology gap with their competitors.

On the brighter side, with the consulting firm having validated the findings of the internal competitive intelligence group, senior management began to recognize this internal resource as a reliable source of intelligence and to utilize their services on a regular and ongoing basis.

Best-in-class organizations recognize that good competitive analysis always results in action or the conscious decision to take no action. It is not a data collection exercise nor an exercise in analysis. It is an exercise in turning data into insights and judgments, and then taking action based on those judgments.

Purpose

Competitive technical intelligence has to have a purpose or it is simply "trivial pursuit"! The purpose has to be defined and driven by the senior management of an organization, because it is at this level that competitive information can usually be acted upon. Actionable information is what it is all about!

A compilation of surveys of companies engaging in CTI cited the reasons listed in Table 2-2, as to why organizations engage in this activity. The reasons most often given are primarily reactive in nature. While most companies operate in a reactive mode, leading edge companies are involved in both reactive and proactive activities.

Avoidance of Being Blindsided.   The most frequently repeated reason given for engaging CTI is to avoid being blindsided. In most organizations, one of the greatest sins to be committed is to allow one's management to be blindsided. Although this is clearly a reactive posture, it may be sufficient justification to have a competitive technical intelligence program in place. In some cases, the information may never be acted upon, but just knowing about events before they become public may help ensure survival within a bureaucracy.

A U.S.-based chemical company was approached by a European firm, offering to license them with technology that impacted directly on the U.S.

Table 2-2   Purpose of Competitive Technical Intelligence

Frequency	Approach	Reasons Given for Undertaking CTI
Most frequently mentioned	Reactive	Avoid being blind-sided
		Acquisition of technology
		Increase external awareness
↓	↓	Component of a tactical plan
		Component of a strategic plan
Least frequently mentioned	Proactive	Input for redeployment of resources

firm's core competency. The company was shocked, because this was the first they knew of the existence of this technology, even though they routinely "monitored" the patent literature. A subsequent recheck of the literature uncovered the patents referring to this technology, but the lack of a CTI system failed to identify these patents initially. The U.S. company was very fortunate that the European firm had no intentions of introducing the technology into the North American market on its own. Subsequently, the U.S. company obtained an exclusive North American license, avoiding a serious competitive threat to its business.

Acquisition of Technology.   Technology acquisition can be either reactive or proactive, depending upon the circumstances. In one instance, it may simply be an opportunistic reaction to a competitor's actions. In another instance, it may be the result of a strategic plan of action, with specific technology acquisition pursued in a proactive fashion. Therefore, one has to look beyond the action itself to understand the driving force.

Increasing External Awareness and Focus.   The first step in the development of a proactive function is to increase external awareness and focus. The intelligence gathered may not be actionable initially. Management's vision is that increasing their organization's awareness of external forces will lead to utilizing this knowledge to build more comprehensive strategic plans. These external forces include political and regulatory climates, global events, economic climate, societal and human developments, and technology trends. Ultimately, this increased external awareness can result in the routine utilization of technology acquisition and strategic alliances to augment internal programs.

Input for a Component of a Tactical Plan of Action.   This initiative can be either reactive or proactive, again depending upon the driving force. Most of the companies surveyed, that engage in CTI, use the information to modify tactical plans that impact primarily on product development and occasionally on the marketing function. Therefore, actions taken are primarily reactive. Tactical use of CTI can be proactive if it is being treated as a component of a strategic planning process, as opposed to being limited to only opportunistic action.

Component of a Strategic Plan.   When CTI is integrated into the strategic plan, its fullest potential can be realized. Redeployment of R&D resources to anticipate future competitive threats or to capitalize on an emerging technology is the payoff in terms of proactive action that can be taken as a result of competitive technical intelligence.

Benefits

Organizations that described their purposes for engaging in competitive technical intelligence also identified the benefits shown in Table 2-3.

Anticipating or Countering Threats.   By anticipating or countering threats, the enterprise can either change the attributes or characteristics of their product, process, or service, resulting in a better design and more competitive position, or develop better market positioning, resulting in a better revenue stream. Establishing more proactive initiatives and anticipating threats can also result in beating the competition to market.

Saving Costs in Development.   A company can speed up product development at lower cost by copying the competition. In the role of fast follower, companies need to track competition very closely, to provide sufficient time to react to new product initiatives. For example, in the consumer goods sector, Dial Corporation has chosen the role of fast follower when competing with much larger companies such as Procter and Gamble, Unilever, or Colgate Palmolive.

Better Positioning of Product, Better Revenues.   Several companies surveyed indicated that the chief benefit of competitive technical intelligence is tactical in nature. CTI enables these companies to add or modify product features on new models and thereby improve their competitive positioning. The result is increased revenues that would not have been realized without the adjustments in features prior to commercial launch. This seems to be particularly true in the automotive and consumer electronics industries.

Motivation Rallies the Troops.   The use of charts, graphs, and matrices to communicate competitive information literally provides a shared view of where the enterprise stands versus its competition. When used to set goals for the organization, competitive technical intelligence acts as a rallying point for the troops, facilitating alignment of their efforts with those of their management. Institutionalizing the competitive technical intelligence func-

Table 2-3 The Value of Competitive Technical Intelligence

1. Allows an organization to:
   - Anticipate threats
   - Counter threats
   - More proactive versus reactive initiatives
2. Save costs in development (use competitors' ideas)
3. Better positioning of product, better revenues
4. Motivation rallies the troops
5. Better focus of the management

tion provides a "scorecard" against which the troops can periodically measure the results of their efforts on an ongoing basis.

Better Focus of the Management.   Institutionalizing the CTI function not only provides a scorecard for the troops but for the management as well. It enables management to assess their enterprise's technology position at any point in time, and provides insights into establishing goals for the organization. By keeping their eyes on competitive technology trends, managers remain focused on future direction, enabling them to anticipate future developments and deal with these developments in a more proactive manner.

Metrics

How does one know when a CTI program is effective? Table 2-4 tabulates how companies surveyed measured the results of their CTI Programs.

Lack of Being Blindsided.   The old adage says that it is impossible to prove a negative. Yet the two top survey responses were, "lack of being blindsided" and "lack of product development failures". In some cases, a CTI program was established after an enterprise was blindsided. It is analogous to installing a burglar alarm system after a burglary, then attributing no further burglaries to the presence of the alarm system. Years ago when my wife and I had a burglar alarm system installed, the installer asked us when were we "hit". When we told him we hadn't been hit, he said that we were very unusual customers, because most installations are made after a burglary has occurred. The point is that lack of being blindsided attributed to a CTI system has to be more of a belief than a metric. The CTI system probably is providing a flow of information, which provides the comfort about not being blindsided. If respondents to the survey had dug deeper, perhaps they might have come up with the number of informational "hits" as a more meaningful metric.

Lack of Product Development Failures.   Similarly, lack of product development failures as a metric probably reflects the experience of companies before and after instituting CTI systems. It is most likely that these companies utilize CTI as a component of tactical and strategic plans to optimize product positioning in the marketplace.

Table 2-4 Measures of Competitive Technical Intelligence Effectiveness

- Lack of being blind-sided.
- Lack of product development failures
- Success stories
- Demand for competitive technology intelligence services

Success Stories.   While it may be a soft metric, success stories or number of hits seem to be the most common positive metric used. Information scientists, whose full-time job is to identify new technologies of potential interest to their organization, commonly use this metric to measure their performance.

Demand for CTI Services.   Measuring the demand for CTI services seems like a convoluted metric. Don't knock it. As described in the case of a lack of credibility, once the value of CTI was demonstrated, senior management's demand for services escalated rapidly. This may be the most powerful metric of all—client satisfaction and continued or increasing demand for CTI services.

Beyond Metrics

There are implications for good CTI beyond metrics, that many practitioners report. These are shown in Table 2-5.

Beating the Competitor to Market.   The CTI program told Company A that its leading competitor, Company B, was developing a new polymer technology that was of importance to Company A's business as well. Further investigation identified Company C, not a direct competitor, who was further along in the same polymer development activity. By acquiring the technology from Company C, Company A was able to beat the leading competitor, Company B, to market, resulting in increased revenues, earnings, and market share at Company B's expense.

Faster Development.   No doubt, technology acquisition or licensing saves development time. In the example cited above, Company C had already been working on the new polymer technology for eight years before Company A requested a license. Where the normal new product development time was five to ten years, commercial launch of products based on the new polymer technology by Company A was effected only three years after the license was obtained. In another example, a joint development program between a pigment supplier and paint manufacturer resulted in the development and com-

Table 2-5 Implications for Good Competitive Technical Intelligence

---

- Beating the competitor to market
- Faster development
- Change in product or service attributes
- Better marketing position
- More competitive cost
- Better design

---

mercialization of a new pigment in only four years compared to the usual ten-year product development cycle.

Change in Product or Service Attributes.   The well-known story of the development of the Ford Taurus automobile is a good example of the use of benchmarking to select specific product features to be incorporated into the vehicle. Surveys of the consumer electronics industry also indicate common use of CTI to modify final model designs prior to launch, based on competitive information.

Better Marketing Position.   Using a CTI study on manufacturing facilities for a specific product line, a company discovered that the industry was operating at full capacity. As a result the company was able to successfully increase prices in what was considered to be a highly price-competitive commodity business.

More Competitive Cost.   In commodity businesses, a critical success factor is to be the low-cost producer. In one such business, characterized by heavy capital investment in manufacturing facilities, a company selected a newer, lower cost manufacturing process developed by a competitor rather than staying with its own proprietary process, when constructing a new manufacturing facility.

Better Design.   Without information regarding competitors' designs, the enterprise runs the risk of suboptimizing its design, resulting in a competitive disadvantage in the marketplace. Some companies surveyed reported rewarding field sales people who were able to provide information on their competitors' design changes prior to commercial launch.

The Plan Is the Driver, CTI Is the Tool!

Upon initiation of a competitive technical intelligence program, organizations can become euphoric about the new information learned, because it corrects previously held opinions, and provides new insights into the competitive scene. This euphoria can result in people becoming enamored with CTI just because "it's nice to know", and it becomes a self-perpetuating function in a way that the original purpose is forgotten. In such cases, emphasis is placed on developing sophisticated databases, and distributing newsletters on competitive findings. Databases and newsletters that are part of an action-oriented program can be very useful, as long as they don't become the end in themselves.

Allowed to become an end in themselves, CTI programs can continue for long periods of time, although they no longer serve a useful purpose. However, in today's climate of downsizing and reorganizations, such luxuries are not likely to be tolerated and become easy targets for elimination. A colleague once asked, "What will it take for competitive technical intelligence to be perpetuated as an integral function within organizations as opposed to becoming a passing fad?" As with any other initiative, CTI needs to be actionable and, in fact, the basis for decisionmaking and subsequent action. Ultimately CTI needs to be incorporated into existing structures and processes within the organization.

To remind the reader, a major reason for engaging in CTI is to identify external technologies that can augment internal industrial research. To facilitate this initiative, strategic and tactical plans, when they exist, become the driving force behind the need for CTI. The plan creates the need. Competitive technical intelligence is the tool to help identify new products, processes, services, and technologies outside the organization. CTI helps meet the growing needs of the enterprise, as defined by the strategic and tactical plans.

GUIDING PRINCIPLES

Principle #3:   Competitive Technical intelligence must be actionable or it becomes "trivial pursuit."

Principle #4:   Senior management must use and promote competitive technical intelligence if it is to become an ongoing activity within the organization.

3

Focusing the Competitive Intelligence Program

Forces Driving Industry Competition

Design of the competitive intelligence search is critical. Who is the competition? Which companies should be analyzed? It is important to define the universe while narrowing the focus of the search. As a reference for considering who might represent current and future competition, a modified version of Michael Porter's classic model of the forces of competition is shown in Figure 3-1.[1]

Often the first impulse is to place one's company in the center box, titled "The Industry", in the belief that competition is coming from similar companies with in-kind technology. This center box may not be where the business belongs. It depends upon the company's relationship to competition. For example, suppliers to the automobile industry offering plastics to replace steel components would be positioned as substitute products when viewing the steel companies. When viewing other suppliers of plastic components, they would then be positioned within the industry.

The same plastics company entering the market for automotive components for the first time would be positioned as a potential entrant. A supplier of bulk plastic resins, integrating forward into supplying molded components, thereby competing with their customers' markets, would be positioned in the suppliers box, while buyers integrating backwards to replace external suppliers belong in the buyers box. The view of who is the competition changes significantly depending upon into which box a company is placed.

In analyzing its competitive position in a niche insecticide business, an agricultural chemicals company was dismayed to find that it was in a lagging

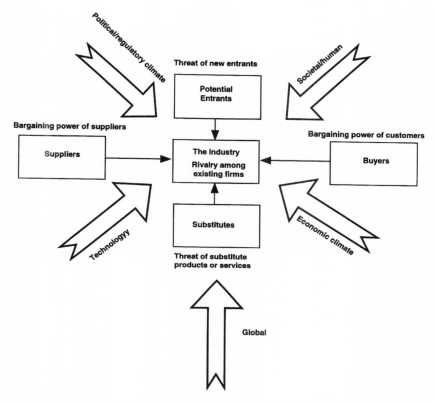

Figure 3-1. Key influences on technology trends, component technologies, and competitive position. Adapted from reference 1.

technological position compared to much larger agrochemical companies. Yet it was the recognized leader in its niche insecticide business. This discrepancy was resolved when it realized it had not focused on its actual competitors, who it dominated, but rather were assessing the industry giants who might represent latent competition, but who posed no immediate threat to its business.

In another example, a new entrant to one segment of the electronics business, utilizing a new paradigm in technology, recognized that the existing competitors could not respond by simply improving on its own existing technology. The incumbents would have to find other companies possessing similar technology to the new technology being employed by the new entrant. By thinking this through before entering the market, the new entrant was able to identify the one company that also possessed this type of new technology, and was able to acquire exclusive rights to it, thereby strengthening its competitive position, and substantially reducing the ability of the incumbents to respond to its new technology initiative.

External Forces

In addition to the competitive forces in the marketplace, key influences in the external environment need to be considered in evaluating technology trends. These external forces can be placed into five classifications; political/regulatory climate, societal/human trends, economic climate, technology trends, and global trends and issues.

A useful exercise is to identify the external forces that will drive technology by working through scenarios looking out into both the strategic and tactical time windows. The exercise involves creating three cases for each force—a best, a worst, and a most likely case. The strategic time window is the amount of time necessary to develop a new clean-sheet design, product, process, or service, from idea generation to commercial launch. The tactical time window is the period of time short of this strategic window.

For example, in the paint industry the strategic time window ranges from five to ten years. Generally this time period holds true for much of the oil and chemical industry. Ever since the passage of the original Clean Air Act in 1972, the primary external driver for the paint industry has been EPA regulation for air and water quality, as well as solid waste disposal. It is not surprising that much of the new coatings technology developed in the United States over the past 25 years was in response to EPA regulations.

In another case, automobile seat manufacturers have sought competitive advantage by anticipating trends in societal/human needs for comfort, as well as increasing regulatory trends regarding safety. An example of this is Acura's advertisement of a car seat, promoting its comfort features, the lack of fatigue, and helping to keep motorists awake and alert when driving.

Companies in the aerospace industry find that they are driven by the economic climate, coupled with global trends and technology.

Segmentation

To repeat, there is so much data out in the public domain, that unless one focuses the search, one can figuratively drown in data! The idea is to collect no more data than will be needed to take action. This assumes that the purpose for the study has been clearly defined with an upfront vision of possible actions resulting from the work. It is therefore important to focus the search into specific segments of interest, thereby limiting the data collected to material that is more likely to be actionable.

Table 3-1 depicts some of the possible segmentation categories.

Business Segmentation.   The concept of business segmentation seems, on the surface, to be self-evident, but getting to the correct level of specificity may take further thought. For example, the automotive paint business may, at first, seem like a satisfactory segment for study. However, further analysis shows that this business can be broken down into the OEM business and

Table 3-1 Segmentation of Competitive Intelligence Effort

Business
Market
Geography
Product/Product Technology
Process/Process Technology
Service/Service Technology
Manufacturing Process/Site

aftermarket segments. These segments represent two distinctly different businesses, and unless they are analyzed separately, the quality of the effort will be adversely affected.

Market Segmentation.   How does market segmentation differ from business segmentation? In the automotive paint example cited previously, the OEM market could be segmented into passenger cars, light-duty trucks, and heavy-duty trucks and trailers. The aftermarket for automotive paints can be segmented into over a dozen components, including such segments as independent body shops, national car repainting chains, and new car dealers.

Geography.   Geographic segmentation is, again, industry-specific in terms of what geographic regions share common characteristics. In the OEM automotive industry, the major geographic markets are North America, South America, Europe, and Asia. Depending upon the nature of the investigation, further geographic segmentation may be necessary. For example, the European market is comprised of a number of diverse markets such as France, United Kingdom, Germany, Spain, and Italy, not to mention the Eastern European countries.

Product, Process, or Service Segmentation.   Products, processes, and services are linked to the strategic business plan as shown in Figure 3-2. A competitive intelligence search can be conducted at any or all three levels of the plan, focusing on customer attributes, the product/process/service concepts, or on the supporting technology.

Customer Attributes.   Customer needs and wants make up the customer attributes. Substantial effort goes into market research to identify which customer attributes, if satisfied, will result in increased revenues and earnings as opposed to those attributes which must be met to simply be competitive. The previously mentioned Acura automotive seat advertisement focuses on two customer attributes which Acura hopes will differentiate its product—comfort and safety. Once an attribute or need is reasonably met by a new product, process, or service, and it is no longer considered to be a "need or want" by the customer, it becomes a "must" in future offerings by the supplier to simply be competitive. Fulfillment of customer musts generally

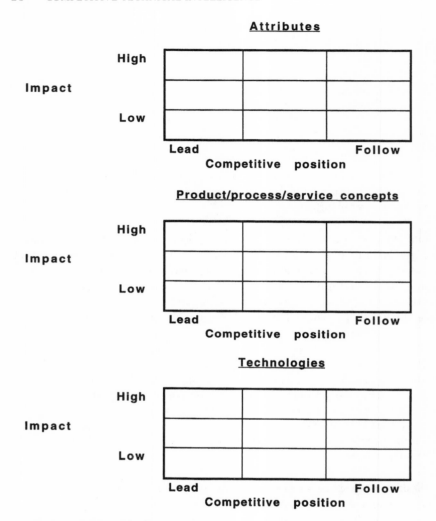

Figure 3-2. Business linkage.

doesn't provide the supplier with an opportunity to increase market share, revenues, or earnings.

A competitive analysis can be undertaken by focusing on attributes that have the potential of differentiating one's offering from those of the competitors. This analysis represents another way to focus the search.

Product/Process/Service Concepts.   In strategic planning, customer attributes, along with external forces, are the drivers to stimulate the generation of new product, process, or service concepts to meet these needs. Another way to focus the competitive analysis is to address these concepts.

Technology Selection for Investigation.   The third link in the strategic plan is technology. Identifying technologies for competitive analysis can be a trying but rewarding experience. The issue is to define a level of abstraction that is relevant to one's activities, yet is searchable in the literature. In the coatings field, examples of technologies would include such diverse items as cross-linking, color matching, and spray application. If one integrates these technologies together into "bundles", we would have core competencies. In the coatings example, a core competency is formulation which could consist of a bundle of technologies including cross-linking, color matching, and spray application, as well as a host of other technologies. Generally, core competencies such as formulation, are not effective search topics because they are too general and will trigger too much irrelevant information.

Underlying the technologies cited above, are more fundamental technologies that are needed for *understanding*, but are again too general to use as a search driver. In the coatings example, a fundamental technology might be polymer chemistry, clearly a topic too broad for effective searching and analysis.

To conduct an effective search, the preferred approach is not to get too concerned about technology definitions, but rather to be "sloppy" and select relevant technologies, product, process, and service concepts, as well as attributes that are relevant to the activity under study. To avoid nomenclature disputes, this mixture can be labeled "selected technologies", even though some of the search terms are not technologies at all.

Manufacturing Process/Site.   In process-intensive industries such as chemicals and oil refining, search protocols can be developed around specific manufacturing processes which may be widely used or in use at one specific site. Focusing on competitive processes enables the enterprise to stay on top of issues such as cost, yield, capacity, and quality.

A Combination of Segments.   In focusing a search, investigators are likely to use a combination of the segments cited in order to further narrow the search field. For example, in the field of automotive coatings, the investigator might limit the search to automotive, OEM, corrosion-resistant primers in North America. Once the search is focused, and the limits defined, the sources of competitive technical intelligence can be considered.

Key Design Considerations

To conclude, the investigator needs to consider six questions in designing a search for competitive technology information:

1. Where does one's organization fit on the competitive forces diagram?

2. What are the relevant external drivers in the universe for the business?
3. Who are the current competitors?
4. Who might be the latent competitors?
5. What segments need to be focused on?
6. Which segments should be integrated into one search and which should be considered sequentially?

GUIDING PRINCIPLE

Principle #5:   In designing the competitive technical intelligence search, consider external influences and latent competition, as well as current competitors within the industry.

4

Sources of Competitive
Technical Intelligence

Conventional Sources of Competitive
Technical Intelligence

Table 4-1 lists many of the conventional sources of competitive technical intelligence. The problem with such lists is that they tend to be random in nature and require further classification if they are to be a framework for a carefully designed search. Some of these sources are of limited value for strategic technology assessment.

Current Literature. For new entrants to a field, the current literature, which consists of texts, journals, and other publications, is a good way to get apprised of the state of the art. However, the current literature is of somewhat limited value for competitive analysis, because on many occasions the products and technologies described are already commercially available or are of little commercial interest. Even academic papers are becoming more limited as an early source of competitive intelligence as universities have become increasingly careful about preempting their ability to patent their discoveries by publishing prior to filing for patents.

This is not to suggest that the current literature be ignored! A U.S.-based chemical company learned of the formation of a European joint program in a technology of interest, by its two leading competitors in Europe, only when it was announced in a publication. The U.S. company was surprised by the news. It is debatable whether the U.S. company could have taken some type of action had it known about the joint program ahead of time, but it would have liked to have had that option.

Table 4-1 Conventional Sources of Technology Intelligence

Literature—state of the art
Patents—new/emerging technology/products & processes
Conferences
Sales/marketing organization—customers & competitors
Purchasing function—supplier contacts
Federal EPA office—products/intermediates (TSCA)
Local EPA Offices—manufacturing
Newspapers in competitors' locality—manufacturing, R&D laboratories, corporate/
    business headquarters
Competitors' products—reverse engineering/analysis, performance characterization

---

The scientific literature can also be a good source of early scientific information published by universities and institutes, or any organization in which scientists are encouraged to publish their results. Such information can be found by routine reading of periodicals, or through literature searches often conducted on the Internet. Searching the Internet will be covered more thoroughly later in this chapter.

Patents.   Actually, patents are an excellent primary source of competitive intelligence when handled properly, as will be discussed later. However, they are of limited value if read routinely as one would read a newspaper, which is a common practice.

A U.S.-based chemical company was approached by a European firm, not active in the United States, with an offer to license new technology which was directly related to the U.S. company's field of interest. The U.S. company felt blindsided! This offer was the first time that the U.S. company had learned of this technology, and the U.S. company wondered how it could have missed it, in spite of its routine monitoring of the patent literature, worldwide. A subsequent review of the patent literature uncovered the relevant references, which had been there all the time. However, because there were only two patent references, it was easy to overlook them by routine weekly reading of patents. A disciplined, focused surveillance program would have readily identified the new technology references.

Conferences.   An excellent primary source of competitive intelligence for specific pieces of information, conferences will not provide the material for a comprehensive assessment. Conferences are of particular value because of the contacts made, as opposed to the papers presented per se. The key is knowing *what* to ask of *whom*, and therefore conferences are a good adjunct to a more comprehensive search program.

Coupled with the "homework" done by reading the relevant scientific literature, one can effectively utilize one's attendance at conferences by making follow-up contact with key scientists.

Sales and Marketing Organizations.   Information fed back from the field is useful in verifying precommercialization projections on technology made from early warning sources. While useful for verification, field information tends to come too late to be of any strategic value, and may or may not allow timely reaction even in the short term.

Purchasing Function.   As with conferences, suppliers can be valuable secondary sources of very specific pieces of competitive information, if a purchasing department knows what to ask and of whom. Like marketing and sales feedback, purchasing's information may come too late for the organization to take tactical, let alone strategic, action, but the information can be a good source of verification of intelligence previously developed from other sources.

EPA Offices.   Federal, state, and local EPA offices are excellent primary sources of specific information for products and intermediates, such as the Toxic Substances Control Act (TSCA), and for manufacturing facilities. It is, again, a matter of knowing what one is looking for. It is not likely that these sources will provide a comprehensive picture, rather specific pieces of information to fill in gaps in data developed from other sources.

Newspapers in Competitors' Locality.   Local newspapers can be particularly useful if monitored on an ongoing basis to track developments at specific sites, such as manufacturing, R&D, and corporate sites. Local newspapers are also a good way to obtain personnel headcounts at locations of interest.

Competitors' Products.   Reverse engineering and performance analysis are common competitive intelligence practices. Unfortunately, these practices are purely reactive after competition has commercialized the product.

Phases of Competitive Intelligence

A useful way to look at technology sources for strategic and tactical assessment is to first consider the extent of development of the technology. Merrill Brenner, of Air Products, depicts the phases of competitive intelligence, as shown in Figure 4-1[1], which can be described as technology search, competitive technical intelligence, and competitive business intelligence.

Technology Search.   The technology search phase focuses on identifying opportunities in new technologies before patents start to be issued. At this point the technology is in the idea/concept stage, and efforts are being directed to determine the feasibility of further pursuit of these leads. The focus of such searches is to identify opportunities in precommercial science and technology. Such searches must be highly focused and specific. The Internet is useful

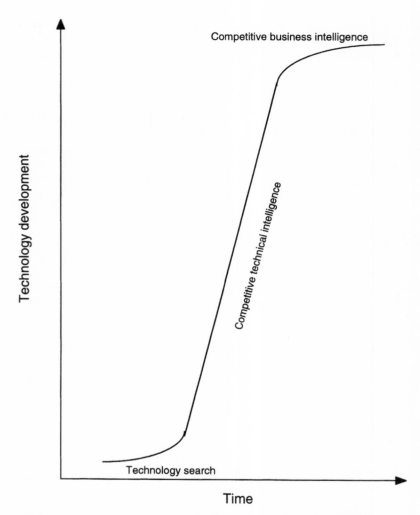

Figure 4-1. Phases of competitive intelligence. Adapted from reference 1.

for such searches; you can investigate publications from both university and government laboratories, as well as small company requests for funding. Another approach is to use co-citation analysis. This approach tends to identify the "players" in a particular field of science or technology. Subsequently, these individuals become the targets for direct contact. Benefits include saving time and generating new options.

Competitive Technical Intelligence.   In this stage, technology development has progressed past concept and feasibility assessment, and development has been initiated, but no commercial activity has yet begun. Patents are being filed and issued. The focus of intelligence is on specific technologies under development by actual or potential competitors. Trends can be measured along with size of efforts. Patents are a prime source of written information. Other key sources are internal and external gatekeepers, lead users, and strategic suppliers. Benefits include avoiding surprises, reducing risk, identifying opportunity, saving time, and enabling planning for proactive response.

Competitive Business Intelligence.   At this stage, the technology is commercial, and is embodied in new products, processes, or services. Intelligence focuses on actual competitors and their products, processes, or services. Emphasis is on market and industry trends, business strengths and weaknesses, strategies, tactics, and costs. Patents, again, are a very relevant source of published information, as well as journals, company publications, and filings with government and regulatory agencies. Other sources include one's own sales organization, customers, and suppliers. Benefits include improved enabling reactive responses, such as pricing strategies, product positioning, and warning of potential or actual obsolescence.

Technology Sources for Strategic and Tactical Assessment

Table 4-2 outlines seven key sources of intelligence that will provide a comprehensive picture as well as sufficient lead time to allow for proactive, strategic, or tactical measures to be taken. More than one of these sources need to be used to verify information developed from any single source.

Patents—Competitor Technology Profiling.   It may be surprising to many, but an excellent way to identify new technologies is to profile one's competitors, particularly those considered to be technology leaders in the industry.

There are several reasons for this observation. First, profiling allows for a highly focused search to safeguard against "drowning in data." Second, one would naturally expect to find the historical technology leaders to be working with the newest technologies.

Patents are the preferred written source of material for profiling competitors in product or process technologies. According to Derwent, one of the

Table 4-2 Technology Sources for Strategic Assessment and Forecasting

1. Patents—competitor technology profiling
2. Scientific literature
3. Conferences
4. Internal gatekeepers—contacts in professional societies, trade associations, and suppliers
5. External gatekeepers—e.g. key expert panels
6. Lead users
7. Strategic suppliers

major abstracters of the worldwide patent literature, eighty percent of new technology information is first revealed in patents. Patent analysis enables the analyst to assess the number of major programs being conducted by a competitor, as well as the competitor's areas of focus, size of efforts, and trends.

Skeptics, critical of the use of patents, justifiably point out that in the United States there is, on the average, an 18-month to 2-year time lag between filing and issuance of patents, so that one is looking at inventions made sometime earlier than the date of the patent's issuance. However, consider how long it takes most industries to move from a clean-sheet design (a totally new development as opposed to an incremental improvement) to commercial launch of a product. In the chemical and petroleum industries this time to market is at least 5 to 10 years. This time period is called the strategic window. Given this time lag, looking at inventions that were made 2 or 3 years prior still allows one to take proactive strategic action before the new technology is likely to appear on the commercial scene.

Frequently, the time lag between invention and publication can be further reduced by searching the foreign patent publications, which issue prior to the patent being granted.

The effectiveness of patent analysis will depend on the availability of a sufficient number of patents to plot trends and differentiate between major and minor efforts. In the chemical and petroleum industries, patent availability is usually not an issue, as shown in Figure 4-2 [2].

As this chart shows, the patent density is high for the chemical and petroleum industries, so there is a strong likelihood of finding enough patents to plot trends in any segment under study. The same likelihood applies for fields such as computers and electronics. Some of the new pharmaceutical areas are undergoing rapid pace of change, but the fields may be so new that a large historical base of patents does not yet exist. Such industries may or may not lend themselves to patent trend analyses, depending upon the rate at which a patent base is being developed. Industries with a low pace of change and sparse patent density, such as construction, tend not to lend themselves to patent trend analysis.

Scientific Literature.   The scientific literature can be a particularly useful source of material from universities and other research institutes, where researchers are encouraged to publish their results early in the development phase. Often, members of the R&D community will identify leads from this source as a result of their habitual reading of scientific publications relevant to their field of interest. Alternatively, the information will be picked up through literature searches, where the Internet can be of particular help.

Conferences.   Conferences afford investigators the opportunity to meet with key members of the technical community in a personal face-to-face fashion. Usually, the atmosphere is a relaxed and social environment, as compared to the workplace. Having done one's homework prior to attending the con-

Pace of change

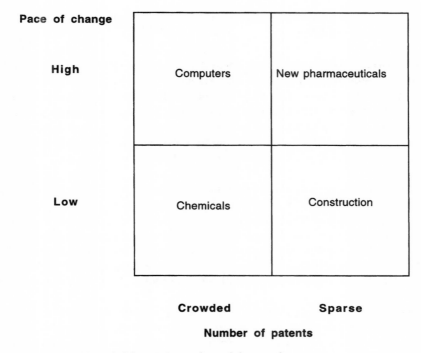

Figure 4-2. Patent field activity. Adapted from reference 2.

ference, the investigator can often flesh out desired competitive information in a low-key, informal setting.

Internal Gatekeepers. An organization's internal gatekeepers represent the "nodes" within the human information network. These individuals make it their personal business to know everything that is going on in their field of interest. They are excellent communicators, are only too willing to share what they know, and are delighted to play an advisory role. Internal gatekeepers can operate synergistically with patent trend analysis. They can help focus the patent search and assist in the interpretation of results. Because of their involvement in society meetings and their tendency to know all the key players in their industry, the gatekeepers can also obtain specific information for the analyst, if the analyst can pinpoint the information gaps in the competitive picture.

External Gatekeepers. Both patent analysis and internal gatekeepers represent internal assessment of competitor activities and external technology trends. At least one external source is needed to validate or challenge the view developed internally. The use of external gatekeepers is a cost-effective way to fulfill this need. A particularly efficient way to utilize these external gatekeepers is through a framework called key expert panels, which is a form of controlled brainstorming. Key expert panels will be discussed in greater detail later in this book.

Lead Users.   In every industry there usually are one or more companies that have the reputation of being the innovators. As a supplier to one of these companies, much can be learned about new technology, product, or process trends as an active participant in a joint development program with a lead user. For example, a supplier to the automobile industry can achieve a stronger competitive position through close development programs with such historical lead users as Daimler Chrysler, BMW, and Honda.

Strategic Suppliers.   In utilizing the patent literature to identify sources of technology, frequently an excellent source turns out to be a supplier to the industry. The patent literature is an excellent way to determine which suppliers are actively engaged in research of interest to one's own organization. The suppliers can subsequently be targeted for potential joint development efforts. Through this sort of collaboration, a strategic supplier can become an excellent source of competitive intelligence, as well.

Selection of Technology Sources.   Figure 4-3[3] places the various sources of competitive technical intelligence into perspective. Each source can be characterized as primary or secondary in terms of proximity to the analyst, as well as personal or impersonal in character. The seven sources just cited cover three out of the four quadrants of the grid.

Patents are a primary source of information and are impersonal in character, as is the scientific literature. Internal gatekeepers are both a personal and primary source, as are lead users and contacts made at conferences. External gatekeepers and strategic suppliers are personal sources in character, but secondary in proximity. None of the recommended sources are found in the fourth quadrant of impersonal, secondary sources, because these sources represent *commercial* information covering products, processes, and services. Sources in this fourth quadrant are fine for getting up to speed on the state of the art within an industry, but are not generally useful in strategic or tactical assessments.

Online Services and the Internet

In a paper presented at a meeting of the Chemical Management and Resources Association, Priscilla Ratliff, of Ashland Chemical, provided a comprehensive assessment of the role of online services and the internet.[4] The following is based on that paper.

Understanding the Internet.   The Internet is basically an international network of computers, which was started about twenty years ago by the U.S. Department of Defense (DOD) to support military research. The concept was

Proximity

	Primary	Secondary
Personal *Character* 	Personal networks • *Internal gatekeepers* Visits Conferences Trade shows Venture capitalists Universities *Lead users* ...	External gatekeepers • *Key expert panels* Consultants Editors Strategic suppliers Analysts Retired executives ...
Impersonal	*Patents* Patent citations Literature Searches Scientific literature Reverse engineering Marketing material Annual reports, 10Ks Ads for staff ...	Industry surveys Trade journals Associations Government studies UN reports Local newspapers Buyers guides ...

Figure 4-3. Sources of competitive intelligence. Adapted from reference 3.

to design computer networks that could continue to operate if parts of the networks were disabled, such as might happen in a military attack. Keeping this original purpose in mind helps to understand how the Internet operates. No one computer is "in charge", rather each computer is a peer of the others. The original network was called ARPAnet, which means the network of ARPA, the Advanced Research Projects Agency of the DOD.

The original users of this network were government employees at military research laboratories and government contractors, who were often researchers in major U.S. universities. It is important to remember who the original Internet users were because it reminds us that the Internet is strong on the types of information provided by government and university sources. Profit-making organizations and corporate America are just beginning to have a presence on the Internet, although this presence has been growing at an exceedingly rapid rate.

Technically speaking, the Internet is amazing because computers of different brands, with different operating systems, programming languages, and software, are able to communicate with each other. All of these communications are accomplished today through telephone lines using an agreed-upon protocol for transmitting and receiving messages.

Online Services.   Just as the U.S. government was developing computer networks for military research, other branches of the government, as well as the private sector and academia, began using computers for data processing, information handling, and publishing support. Some of the earliest online computer files began in the mid-1960s. In her paper published by the Chemical Management & Resources Association, Priscilla Ratliff, of Ashland Chemical, identified six different types of available online services. These are tabulated in Table 4-3.

Many of these online services are available on the Internet and more are being added monthly. Therefore, the Internet is simply another way of accessing these services. In some instances, Internet access will be the only practical method, especially in the case of non-U.S. services. One advantage of using the Internet is that only one telephone call is required. Some heavy users may call up and connect in the morning and stay connected all day.

The professional and consumer online services started independently and had access methods whereby one could dial into the host computer directly. Recently professional and consumer online services have joined the Internet as a convenience for their customers so that with one connection a customer may use various services without having to disconnect and redial a different telephone number. Likewise, the specialty online services have also joined the Internet both as a convenience to their customers and to attract new customers. Once the designated service is reached via the Internet it then operates the same as it does via other access methods.

Online Services versus the Internet.   Priscilla Ratliff points out that information on the Internet is not easy for the inexperienced online searcher to access or analyze. At the time of this writing, the Internet is not for the novice online searcher. It is confusing, inconsistent, obscure, time-consuming, and difficult. It has not reached the point where it is a cost-effective tool for daily searching for business or technical information. It also has the added disadvantage of potential lack of privacy or confidentiality for the searcher.

The professional online services provide organization and structure to business and technical information, as well as security and privacy to their customers. In regard to patent searching, they have user-friendly, easy-to-use search protocols. With these search protocols, they are improving their services by offering software routines that aid the professional searcher in further analyzing the results of queries.

New Technology Searches.   Even with the difficulties associated with using the Internet, a particular value has been in searching for technology before any significant number of patents have issued. As pointed out by Merrill Brenner of Air Products and Chemicals, "the first signals often emerge in

Table 4-3 Types of Online Services

Type of Service	Description
Professional level online services, such as DIALOG, STN International, LEXIS-NEXIS, OCLC, DATATIMES, NEWSNET, DATA-STAR, OR-BIT*QUESTEL, DOW JONES NEWS/RETRIEVAL.	Very large online services, which collectively contain thousands of databases of information on all kinds of services. They require special training to use and a considerable investment of time and effort to learn to search proficiently. They are expensive and are basically designed for use by a person in the context of his or her employment or profession.
Consumer online services, such as COMPUSERVE, PRODIGY, and AMERICA ONLINE.	These services compete for the mass market, retail consumer or home user. They are menu or mouse driven. They are priced to be affordable by the user.
U.S. Government online services, such as the National Library of Medicine (NLM) and FEDWORLD.	These services vary quite a bit in user friendliness. They are inexpensive or free depending upon the agency supporting them.
University computer centers, often offer some services to the local community where they are situated.	Services are free to faculty, staff or students and are either free or available at nominal cost to local residents. Some services have restricted access.
Public online services called freenets, which often include local public library catalogs, community calendars, local community services, and local government information.	These services are free but may require user registration. They are menu driven and designed for ease of use.
Specialty online services which are directly available from the producer of the information.	There are hundreds of these types of services which often require some type of subscription or yearly access fee. They may or may not require special training to use and have a fairly narrow focus.

Source: Adapted from reference 4.

scientific and technical discussions, 'gray literature', or statements that resources are being directed to certain areas of science or technology. These signals might be weak, but gathering, assessing and communicating this information are crucial objectives of external technology scouting programs that uncover and anticipate pre-commercial developments."[1] Examples include monitoring small business innovation research grants, as well scientific publications from university and government sources.

Internet sources would be classified as impersonal in character and could be either primary or secondary in proximity, depending upon the nature of the publication.

The issue of focusing the search becomes even more critical when utilizing the Internet. Given the incredible quantity of information available, great precision in the use of key words is necessary to minimize the amount of "garbage" or "noise" that the search collects. Nevertheless, the convenience of searching on one's computer screen is compelling reason enough to include the Internet as a key source in the search protocol. By the time this book is published, the Internet will probably have become substantially user-friendly to the amateur investigator.

GUIDING PRINCIPLE

Principle #6:   To ensure accuracy and completeness, always use more than one source of information.

5

Patents as a Source of Competitive Intelligence

Levels of Patent Profiling

An effective method of patent profiling is to conduct it at three sequential levels, as shown in Table 5-1, starting with Level 1 and moving on through levels 2 and 3. Levels 1 and 2 represent competitor profiling, while level 3 seeks to identify other sources for the technology of interest.

Level 1: Patent Survey

Targeted Segment. At level 1, the idea is to collect all relevant patents in the segment of interest over the time span to be studied. For example, if the targeted segment were coatings for automotive original equipment, key words such as automotive coatings, automotive paints, and automotive OEM coatings might be used along with the name of the targeted competitor. The goal at this point is not to "fish" for a sampling of patents in the "lake", but to drain the "lake" and collect everything. That is, again, why it is so important to have sufficiently focused the search before starting to collect patents.

Level 1 represents a view from "10,000 feet", so it is not necessary to read the claims imbedded in the patent. The search can be limited to a quick review of title and abstract. The patents collected should be organized by date filed, rather than date issued, in order to get closer to the actual date of invention. In addition, it is worthwhile to look at the sum total of patents issued over the time period, to assess the size of a competitor's effort.

Table 5-1 Levels of Patent Profiling

Level 1	Competitor patent survey
Level 2	Selected competitor technology profiles
Level 3	Technology search—extend to identify other sources of a given technology

Strategic Time Window.   As a minimum, the time frame over which patents are collected should reflect the strategic time window for the industry. This strategic line window is defined as the time to go from idea to commercial launch for a 100 percent clean-sheet design. In the oil, chemical, and allied industries, this time frame is usually at least 5 to 10 years. Therefore, in these industries, patent collections should span at least 10 years.

Deliverables.   A level 1 patent survey can provide significant information. Even at this macro level, the size of a competitor's technical effort can be determined by the number of patents and, with a few assumptions, the number of people. With further extrapolation of the cost per assignable researcher, manpower estimates can be converted into R&D dollars for further comparisons, such as R&D as a percent of sales. In addition, by plotting the number of patents filed each year over the strategic time window, the trend for the effort can be determined, whether it is increasing, flat, or declining. As a result, patent surveys profiling a given company can provide the analyst with a reasonably accurate picture of the size and direction of the R&D effort for one or more competitors within a selected segment of the business.

A first approximation of manpower is to count the number of inventors by year, as well as over the time period being studied. This assessment is still superficial, because for each inventor there are usually a number of other technical people working on the project. To go beyond the first approximation, it is necessary to benchmark one's own organization. From this benchmark, a ratio of inventors to total assignables on a project can be developed. Although there is no assurance that a competitor will have a similar ratio, experience has shown that a reasonable estimate in a correct order of magnitude can be achieved. With additional extrapolation, applying one's own organization's cost/assignable to a competitor's headcount can provide a rough competitor cost estimate, again, accurate within an order of magnitude.

Clearly, such approximations need to verified through other sources before they can be accepted as accurate intelligence. At the least, this methodology serves as a starting point, and provides insights into questions to ask in pursuing verification. Even if subsequent information proves these assumptions to be incorrect, the amount of time taken to have produced the extrapolations is minimal.

Level 2: Selected Technology Profiles

Identification of Selected Technologies.   At the second level of patent pro-filing, competitor activity in selected technologies of interest is analyzed. Identifying technologies at the correct level of abstraction is a challenge. The term "selected technologies" was deliberately chosen to avoid debates about definitions of core technologies versus product or process concepts. The technology has to be at an appropriate level of abstraction so that key words can be identified that are searchable in the patent databases. For example, the term "polymer chemistry" is at too high a level of abstraction, while the term "condensation polymerization" would be more specific and, therefore, more searchable. For search purposes it is acceptable to mix products and processes with technologies, for example "water-based paints" (a product concept) is an acceptable search protocol to include along with technologies.

Sources of Selected Technologies.   An ideal source of selected technologies is the organization's strategic technology plan, if one exists. A strategic tech-nology plan should already contain the technologies of interest and usually describes them in actionable terms at a level of abstraction suitable for search protocols.

In the absence of a strategic technology plan a list of selected technologies can be developed through interviews with technical management and R&D gatekeepers. To ensure organizational "ownership" for the technology list, it needs to be reviewed, upgraded, and refined with senior management, preferably with all relevant functional stakeholders in addition to R&D, usu-ally marketing, and manufacturing.

How many selected technologies are reasonable for search purposes? The number can vary substantially from 10 or so for a modest search effort to many times that number for more sophisticated programs. Regardless of the ultimate number of selected technologies, the reader is encouraged to keep the search simple initially by limiting the number of selected technologies to a level that can be managed manually, such as 10 to 20 technologies. Once the search protocol is refined, additional technologies can be added, and such techniques as computer-aided sorting and classification can be em-ployed.

There are several quick and useful techniques to winnow down the list. One method is to do a qualitative technology impact assessment by rating each technology *high* or *low* in regard to *impact on the business.* High-impact technologies result in increased revenues or earnings, while low-impact tech-nologies help maintain the current position with respect to sales and earn-ings. The rationale here is to focus technology assessment on the higher impact technologies.

An alternative method is to narrow down the segment to be studied. This could allow searching of all technologies of interest in that segment, regard-less of perceived impact, but requires subsequent searches of the segments previously eliminated. The two methods are not mutually exclusive and are often used together to obtain a more manageable list.

To Begin, Keep It Simple!

When initiating a level 1 and the subsequent level 2 search, keep it simple by initially selecting only *one* company to profile. Depending upon the number of patents anticipated to be found, one could limit the initial search further to consider U.S. patents. This search could then be expanded sequentially to include foreign patents and foreign filings. Experienced searchers can include foreign patents in the initial search.

One should expect to make adjustments in the search protocol in case the search is producing too much irrelevant data or, conversely, too little data. Once the search protocol has been adjusted, and the searcher is comfortable with both the quality and quantity of "hits", additional competitors can be added, and the search can be extended to include foreign patent filings as well.

Deliverables

In both levels 1 and 2 searches, the analyst can make determinations regarding sizes and directions of efforts. As a result of the level 2 technology assessments, one can assess one's own organization's current position, as well as predict future technology positions relative to the competitors studied.

Additional information can be gleaned regarding competitors' activity, such as an assessment of the degree to which research activities are focused or scattered. Trends can be extrapolated to predict competitors' potential new product or process initiatives. In a defensive mode, the organization may identify opportunities for future work to meet or follow competition.

Competitor profiling is an excellent means of identifying new technologies. These new technologies usually show up as a small, but increasing, number of patents within a selected technology category. When such a hit is made, the usual next question is, "who else in the world is working in this field?"

Level 3: Other Sources of New Technology

At this point in the search process, a patent search initiative can be undertaken using the newly identified technology as the key word. In such a search, a large number of citations are normally found in a scattered pattern, with only a few citations belonging to one organization. Most of these citations are likely not to be relevant to the technology of interest. If the searcher is fortunate, one or more references will be relevant, thereby identifying one or more alternative sources of the technology of interest!

The implications behind one or more alternative sources can be significant. Do these findings indicate a loss in competitive position for the searcher's organization? Does it signal one or more new entrants into the field? What role might these new entrants play—supplier, customer, or competitor?

What opportunities do these alternative sources present? Do any of these sources represent an opportunity for a strategic alliance? Is there a licensing-in opportunity? Is there an opportunity for a joint development? Do these alternative sources represent an opportunity for an acquisition, or should the investigating organization consider redeploying its own internal effort to meet this threat?

Performance Polymers, Inc.—An Example of Levels 1 and 2 Analyses

Let's see how the three-level search methodology works with the fictional company Performance Polymers, Inc., a diversified manufacturer of engineering polymers. The company is divided into six strategic business units, as shown in Table 5-2.

The planning division has been asked to help each business unit to develop a strategic technology plan. As an initial step, they have chosen to focus on the reinforced injection moldings (RIM) business and to develop a competitive assessment of RIM technology. Having some previous experience with this segment, planning decided to chart all key competitors, rather than starting with only one.

Level 1: Patent Survey.   Table 5-3 displays the key competitors in the RIM business. In a level 1 survey, the key words *reinforced injection-molded plastic* and variations of those words, along with each company name, can be used to determine the number of patents issued per year. It is important to make some cursory assessment of the nature of the patent. Some organizations will take an invention and break it down into a number of patent applications, while others will file a single blockbuster patent. In order to compare "apples to apples", it's useful to recombine continuation in parts and

Table 5-2 Strategic Business Units in Performance Polymers, Inc.

1. RIM (reinforced injection moldings)
2. Foams
3. Elastomers
4. Polycarbonates
5. Nylon
6. Polyesters

Table 5-3 Level 1: Patent Survey Tabulation RIM Business

	Number of Basic Patents				
Year filed	Performance Polymers	BASF	Bayer	Dow	ICI
1996					
1995					
1994					
1993					
1992					
1991					
1990					
1989					
1988					
1987					
Total 1987–1996					

not to double count foreign equivalents. Initially, the study can be limited to U.S. patents, or to an organization's home country. In a second step, the study can be extended to include filings in other countries.

Alternatively, foreign filings can be included in the initial search if one is comfortable with the search protocol. In this case, it may be more productive to initially check filings in Germany as well as U.S. patents for BASF and Bayer, and in the United Kingdom for ICI. In the United States, only issued patents will be found, while patent filings, as well as issuances, will be found in countries outside of the United States. As a result, many disclosures usually show up in foreign filings before their U.S. patent equivalent issues.

Referring to Table 5-3, the number of basic patents for each company, *including the investigating company*, is tabulated over the past ten years. Not all patent filings and U.S. issuances will be captured for the most recent years. This gross tabulation provides the analyst with an assessment of the patent population to be dealt with, so that the segmentation can be enlarged or narrowed depending on the size of the population. If the population is large enough to be plotted graphically, yet small enough to be manipulated with the resources available, either manually or by computer, the analysis can proceed. The sum of all patents for each company is a measure of the size of its respective R&D efforts in RIM. As in warfare, ultimately, size and numbers win. The same can be said in the "technology war" as measured by the number of basic patents.

Table 5-4 RIM Technology Tabulation

Technology	Performance Polymers	BASF	Bayer	Dow	ICI
Polyols					
Additives	Tabulate, by company, the total number of basic patents and inventors in each selected technology for the strategic time period.				
Cross-linkers					

Having tabulated patents by year, a graph can be constructed showing the number of basic patents per year. Remember to classify by year filed, rather than year issued. To smooth out the curves, plot the numbers using a two-year running average.

Because U.S. patents require the listing of all inventors, an estimate of the size and trend of the effort can be made by counting inventors. When tracking inventors be sure not to repeat the same name in a given year nor repeat a name when doing an overall total.

Level 2: Selected Technology Profiles. For illustrative purposes, an oversimplification of the selected technologies is shown in Table 5-4, limiting the technologies of interest to *polyols, additives*, and *cross-linkers*. Unlike a level 1 survey, a level 2 analysis requires several tabulations. The first, Table 5-4, tabulates the total number of patents for each technology by company for the ten year period. This tabulation provides the analyst with a more detailed picture of where each competitor is focusing its respective R&D efforts, and the order of magnitude of that effort.

The second tabulation, shown in Table 5-5, plots the number of basic patents and inventors by year for each company, and for each individual selected technology. These data can then be plotted on a graph to show trends. It is from these trends that a new cross-linker, for example, might be

Table 5-5 RIM Technology Trend Plot Cross-Linkers

Company	1996	1995	1994	1993	1992	1991	1990	1989	1988	1987
Performance Polymers										
BASF										
Bayer	Plot number of basic patents and number of inventors for each company, by year.									
Dow										
ICI										

first picked up. To pick up a new cross-linker a second classification would be necessary, in which the cross-linkers are divided into classes, and then the data plotted in a fashion similar to Table 5-5.

Value of a Level 2 Technology Trend Analysis

Company A had licensed-in technology from Company B in 1990. Company A had not been utilizing patent trend analysis, but tended to follow the patent literature by relying on ad hoc reading of the weekly *Official Gazette* and/or *Derwent Abstracts*. In its desire to test the patent trend analysis methodology, Company A wanted to know how much earlier it might have become aware of the dominating technology of Company B, so that it could have licensed the technology sooner.

Figure 5-1 is a graph showing U.S. patent issuances for three companies. Company A got involved with technology X relatively late compared to Company B, the leader in this particular technology. The graph also shows that Company C, in the role of follower to Company B, in fact had circumvented Company B's patents. The curves have been smoothed through two-year averaging.

From these graphs, Company B's activity starts to become apparent in 1979 with the issuance of about 7 patents in technology X, peaking at about 32 patents in 1985 and tapering off to about 8 patents in 1991. Company A, also working in this field at a fairly low level, producing less than 2 patents per year between 1980 and 1986, increased its efforts in 1987 to about 5 patents, with a peak of 6 patents in 1988, before tapering back to 2 patents in 1991.

Company A became aware of Company B's dominance in 1987, after having increased its own efforts in the field, and, after some negotiations, having taken a license in 1990. Had Company A been doing patent trend analyses, it probably would have picked up on Company B's surging activity in 1981 or 1982. Assuming the same time lag to negotiate a license, Company A could have obtained a license in 1985, rather than in 1990, five years earlier! In addition, Company A could have saved its own substantial R&D effort that probably started around 1985, and could have deployed its resources elsewhere. Furthermore, Company A would have been aware of Company C's activity, which might have been an alternative source for this technology!

To conclude, this illustration points out that regarding technology X, a level 2 technology profile trend analysis could have saved Company A five years in time, several million dollars in R&D expense, the opportunity to be focusing on more productive activities, and the option of considering an alternative approach to technology X as practiced by Company C. As a direct result of this learning experience, Company A has established an ongoing patent competitive intelligence system.

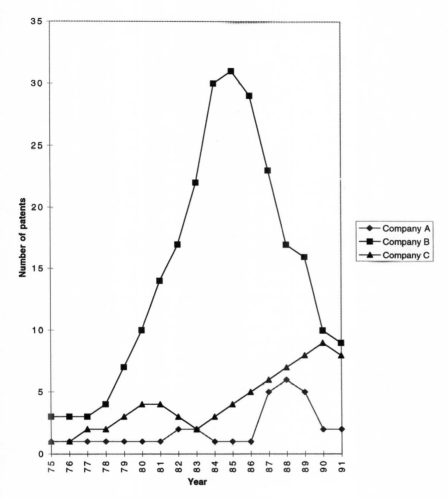

Figure 5-1. Technology "X" patent trends.

An Example of a Level 3 Analysis: Identification of Other Sources

A supplier of paints to the automobile manufacturers was seeking a new technology to improve the resistance of its paint to the etching effects of acid rain, commonly experienced in the United States. A level 2 search identified a new technology, Y, that a competitor was developing and that had the potential to provide the desired level of acid rain resistance. Knowing that its direct competitor would be unwilling to license the technology, the company conducted a level 3 search in the patent literature to determine if any-

Table 5-6 Level 3: Search for Technology "Y"

Company	Number of Patents 1986–1995
A	2
B	3
C	2
D	**12**
E	1
F	4
G	2
H	3
Leading competitor	**9**

one else in the world might be working with technology Y. Using Y as the key word, a search produced the results shown in Table 5-6.

Many of the patent citations found were not relevant to the specific topic as exemplified by Companies A through C and E through H. Yet this study produced an exciting hit! Company D, an Asian-based company, had been actively developing Technology Y for a number of years and, in fact, had more patents than the U.S. company's leading competitor. Furthermore, Company D represented a potential new entrant to the industry. Surveying existing competition worldwide would not have uncovered this company. Subsequently, Company D was persuaded to enter into a successful strategic alliance with the U.S.-based company, and jointly continued to develop the technology into a commercial product, meeting the original goals that initiated the search in the first place.

Other Sources of Competitive Intelligence

In the various examples cited earlier, other sources of competitive intelligence were utilized in concert with patent trend analysis to develop an integrated picture. In the case of the Performance Polymers example, internal gatekeepers provided the list of companies active in the RIM field, as well as the list of selected technologies for study. Internal gatekeepers were also involved in the level 2 analysis of technology X, providing the historical record of events plus the subsequent analysis of Company C. As a result, it was determined that Company C had in fact circumvented Company B's patent claims. Similarly, in the case of the search for technology Y, internal gatekeepers helped design the search protocol and made the judgment that technology Y could be used to improve acid rain resistance. This judgment triggered the subsequent action leading to the development of the strategic alliance with Company D.

The importance of internal gatekeepers was illustrated in all three examples cited above. Their role in competitive technical intelligence, therefore, bears further examination.

GUIDING PRINCIPLES

Principle #7:   A cost-effective way to initiate a patent search for new technology is to first profile competitors' patents.

Principle #8:   Patent trend analyses provide information not only on the size of competitors' efforts but also show whether the trend is increasing, decreasing, or flat.

Principle #9:   Patent trend analyses not only profile competitors' technologies, but also identify the emergence of new technologies.

Principle #10: Utilizing key words derived from the new technology, one can answer the question, "Who else in the world is working with this technology?"

6

Scientific Literature and Conferences

Value of Technology Search via the Scientific Literature

While the scientific literature can be both a source of competitive technical intelligence, as well as a vehicle for searching new technology, its value is primarily in the latter—that is, identifying new technology of interest.

There are several reasons for this observation. First, the most valuable material tends to be published by universities and research institutes, as opposed to the large private industrial firms. These papers usually describe scientific research, as opposed to downstream product or process development. Companies seeking such technology, in the very early stages of development, need to pay attention to these scientific publications.

Second, industrial companies do not tend to publish this sort of work until the technology becomes a commercial entity, or is judged to be of scientific interest, but with no commercial value. Potentially valuable technology, developed by industrial concerns, will first be described in written form in patent applications.

Therefore, information developed by the investigator through the scientific literature tends to fall into the category of *technology search* as described in Chapter 4. Another value of searching the scientific literature is that it is an excellent means of identifying the researchers within the global scientific community who are working in areas of similar interests.

Some examples of utilizing this source for competitive advantage involve activities of scientists within the former Soviet Union. Technology within Russia and the other former states within the Soviet Union had been applied primarily to military applications. Little industrial application was developed. Yet, much of the research being carried out at the universities and research institutes was of world-class caliber, as revealed in the scientific literature.

One of these areas is polymer technology. Initially, key Russian scientists were identified through their publications in scientific journals. Subsequently, one-on-one relationships were developed by western scientists through correspondence, meetings at conferences, and visitations. These activities ultimately resulted in the transfer of scientific information to western companies, including such technologies as new polymerization mechanisms and polymerization catalysts. Western companies with such longer range vision consequently benefited from these liaisons and gained competitive advantage through superior technology.

Staying on top of developments requires several approaches to dealing with the literature. First, motivated researchers have their favorite journals, which they read on a regular basis to stay abreast of new developments. Second, literature scientists can set up selective disseminations of information, (SDI's) in which articles, or abstracts of articles, in specific fields of interest are routinely collected and distributed to interested parties on a periodic basis. Third, specific literature searches can be initiated on a project basis. It should not be surprising that the internal gatekeepers (described in Chapter 7), are often the heaviest users of journals and the SDI's.

Division A in a multidivisional industrial North American chemical company had developed a joint program with an Australian research institute, and was circulating literature regarding the institute to its sister divisions. A researcher in Division B, seeing this literature, recalled some publication by this institute regarding some new catalyst technology. A subsequent literature search verified the researcher's recollection. This led to a dialog between Division B and the Australian institute, ultimately resulting in a second joint development program in this specific field of mutual interest.

Government Laboratories

According to the National Technology Transfer Center (NTTC) at Wheeling Jesuit College in Wheeling, West Virginia, organizations can gain access to $36 billion worth of federally funded research at more than 700 federal laboratories and projects at universities. The NTTC is funded through NASA and other federal agencies to help U.S. businesses become more internationally competitive through commercialization of federally developed technologies.

Access to this technology can begin with a telephone call to NTTC's National Gateway at (800) 678-6882. This is a free service that provides business and industry with free leads to technology, expertise, and facilities. In addition, NTTC operates Business Gold, an electronic bulletin board service, to provide a way for companies to directly access information through NTTC's World Wide Web Home Page at URL http://www.nttc.edu.

A U.S. manufacturer of automobile components was seeking a source of technology to produce low-noise worm screw drives. The manufacturer had identified several sources of technology and products in Japan, and wondered if there might be a U.S. source of such technology. The NTTC was able to provide several leads for federally funded technologies originally developed for use with helicopters.

Conferences

Conferences are an ideal place to meet with one's opposite numbers from other industrial companies, as well as colleagues from academia, research institutes, and government laboratories. In addition to the information provided by the papers presented in the conference, the relaxed, informal atmosphere is conducive for gathering vital competitive information during the informal conversations that take place during the networking process. Conferences provide an opportunity for personal contacts to access primary sources of information.

It is a matter of knowing what to look for, so you must do your homework before attending a conference. Participants in a structured competitive intelligence program within their organization may get "assignments" in terms of the competitive information needed. Because competitive assessment is similar to putting a jigsaw puzzle together, sometimes a relatively innocuous piece of information can complete a competitive assessment. Because the desired piece of information can be innocuous, it often can be readily obtained during casual conversations. Even when specific needs cannot be identified ahead of time, groups of attendees can be organized to cover different segments of trade shows or to meet with specific attendees at a conference. Subsequently, one "coordinator" debriefs the attendees and integrates the information to assess what has been learned. Learning experiences have to be rewarded for positive reinforcement. While this is a common practice among Asian and European organizations, it is not yet a well accepted part of U.S. business culture.

Having been trained in a company that stresses the protection of proprietary information, it always seemed surprising to me how many technical people are willing to discuss their work with only the slightest encouragement. The explanation offered is that many technical people are delighted to find someone interested in hearing about their work, and feel that the information they are divulging is relatively innocuous.

Understanding the Stages of Technology Development

A chronic problem within many industrial organizations is developing a uniform understanding of what stage a newly identified technology may be at, and what will be required in terms of time and effort to develop commercially viable products, processes, and services. The business division may develop an overly optimistic view of a potential technology acquisition, which might have been fueled by the enthusiasm generated within the technical organization.

Negotiations for a joint development proceeded slowly between Division B of the North American chemical company and the Australian institute described in the example cited earlier in this chapter. The institute was afraid that its potential industrial partner would require it to subvert its research into short-term product and process development activities. At the same time, the institute was reluctant to grant exclusive rights to the industrial firm for fear that the firm would "sit" on the technology rather than use it as a driver for commercial products.

The industrial firm was required to carefully lay out a time line describing what would be involved to take the technology from concept to commercial launch of the anticipated products, as well as the technical and commercial risks associated with the planned commercialization process. Once this time line was clearly understood, and expectations between the two parties were in alignment, the joint cooperation program was put in place.

Similarly, it would be unrealistic to think that technology developed at a government laboratory would be a simple "drop in" to an industrial application. Much of this technology may have been developed for defense or aerospace purposes, where cost was secondary, and a substantial effort may still be required to develop a commercially viable product or process for use in the private sector. In working with Russian scientists, it would also have been unrealistic for western organizations to expect the Russians to have near-commercial prototype products or processes.

GUIDING PRINCIPLES

Principle #11: The scientific literature and federally funded research can be excellent sources for early stage technologies.

Principle #12: Beyond the obvious value of the formal programs, conferences can be excellent sources of competitive technical intelligence through informal networking interactions.

Principle #13: It is important to have realistic expectations regarding the time and effort requirements, as well as the risks associated with the commercial development of early-stage technology.

7

Internal Gatekeepers as a Source of Competitive Intelligence

Identifying the Gatekeepers

Internal gatekeepers are the "nodes" in the informal internal human communications network. They are usually top-notch researchers, and know just about everything that is going on in their respective fields of interest. Gatekeepers are involved in the professional world around them, tend to be prolific readers, keep up with the journals in their field, attend relevant professional society meetings, and maintain frequent informal interactions with colleagues inside, as well as outside, their respective organizations. Many gatekeepers are so active that they develop deep interests in fields totally outside their main areas of endeavor. Although rare, there are Ph.D. chemists who have also received advanced degrees in such diverse fields as archaeology and psychology.

If you find it difficult to identify gatekeepers in your organization, one way would be to examine how informal communications flow within the operation. If you can identify the "nodes", you have usually found the gatekeepers.

One of our laboratories was in the process of modernization; the top floor in the laboratory was already modernized and three more floors were in process. Joe, one of our most productive technical people, occupied one of the older labs. To recognize his many contributions, we transferred Joe to one of the modernized labs up in the corner of the fourth floor. Within a week, Joe requested to be moved back to his old lab. It turned out that Joe was a key communicator within the lab. His old lab was located in the middle of the third floor, in a high traffic area. You could usually find one or more visitors with Joe. By moving Joe to the quiet corner of the fourth floor, we had taken him out of the mainstream flow of traffic, and his level of interaction with others had fallen off substantially. We quickly responded to Joe's request, and moved him back into his old lab. Without thinking, we had done a great disservice to one of our top gatekeepers.

Gatekeepers are totally self-motivated. They march to the beat of their own drum. Although independent, they are extremely helpful and enjoy their roles as advisers or consultants to others. Clearly, gatekeepers would have to be an excellent source of competitive technical intelligence as well as a resource for search design and subsequent analysis of the resulting information. Their independent nature raises the question of how to focus their attention on the issues at hand, and to get their cooperation and participation in the competitive intelligence process.

Roles in Competitive Intelligence

The challenge with internal gatekeepers is to integrate their freewheeling, ad hoc style into an organized competitive technical intelligence system. I tried a number of approaches to this issue, so it may be worthwhile to start with what I learned not to do.

First, don't try to assign gatekeeping roles to non-gatekeepers. Gatekeepers are born, not made. They are naturals, and they self-select their roles. A gatekeeper can be readily identified just by observing how technical people interact with each other within an organization, as previously described. Identify and work with proven gatekeepers. It is not a role that can be arbitrarily assigned to an individual just because that person is working in an area of interest to the competitive intelligence investigators.

Second, don't try to give a gatekeeper a competitive intelligence "assignment". Gatekeepers are independent. They operate the way they do because it is their natural style, and they don't expect to be held accountable for their operational style. As soon as they feel they are being given an "extracurricular assignment" for which they will be held accountable, their tendency is to rebel. When we tried to assign formal responsibilities to the gatekeepers, they responded with a number of questions including, "What's in it for me? Will this become part of our position description? Will this become part of how our performance will be judged? Will this result in additional financial compensation or promotional opportunities?"

Nevertheless, internal gatekeepers are achievement-oriented and have a strong desire to have a major impact on the businesses they support and on the supporting R&D programs. This motivation is the key to directing their efforts into something more than ad hoc participation. Their buy-in is essential. An ideal approach is to involve the gatekeepers in the strategic technology planning process and in the subsequent strategic technology plan (if such processes and plans exist). In this activity, you can negotiate with the gatekeepers to accept a leadership role in taking responsibility for one of the selected technologies.

Taking responsibility means actively participating in the strategic technology planning process, and providing the information needed about the specific technology, including the organization's competitive position, as well as the competitive position and activities of others. The gatekeeper should provide a holistic view, as described in the Michael Porter's model of the competitive environment, including forecasting future technology directions. This participation naturally leads to being called upon as a resource for the business. The gatekeeper will rapidly become aware of the information gaps relative to the selected technology being supported, and will subsequently utilize the available resources and networks to fill in the information gaps. Gatekeepers are often, the more senior members of the technical community. If the gatekeeping process is to be institutionalized, successors have to be identified and developed. This can be best done by establishing small teams rather than depending on one individual per selected technology. To be successful, there has to be mutual agreement between a potential team member and the gatekeeper to form a team.

In the absence of a formal strategic technology planning process, internal gatekeepers can still become involved in whatever competitive technical intelligence system exists. The guiding principles remain the same: identify the natural gatekeepers, encourage them to self-select technologies for which they will be responsible, allow them to see how they can have a positive impact on the businesses they support and on the supporting technical programs, and enable them to do just that. Maintaining an effective internal gatekeeper network requires ongoing support and nurturance, as well as a defined framework in which to operate. Failing this framework any system will deteriorate, and the Gatekeepers will revert back to operating on a less effective, informal, ad hoc basis.

What do you do if no gatekeeper exists in an important technology area? Arbitrary appointment, simply because an individual is working in that area, will not work. One solution may be to transfer a known gatekeeper into that area. On a longer term basis, R&D management may need to review staffing in their organization to determine if there is a balance of functions in each important area, such as not only gatekeepers, but also other functions such as entrepreneurs, "project managers", and supportive personalities. Unfortunately, most managers do not tend to view their organizations in these terms, and by default, may rely on serendipity for the gatekeeping function, if it exists at all.

Specific Responsibilities

I previously indicated that one cannot unilaterally hand out an assignment to an internal gatekeeper without inciting a revolution. Alternatively, as the

Table 7-1 Suggested Responsibilities for Internal Gatekeepers

---

Keep decision makers up to date on the state of the art of the selected technology.

Provide an understanding of the organization's competitive position with regard to their selected technology.

Pinpoint future selected technologies for competitive profiling.

Participate in the competitor profiling process.

Participate in technology forecasting.

Help assess impact of future technology options.

Help draw conclusions and recommend strategies.

Train future internal gatekeepers.

---

gatekeeper buys in to the strategic plan or the competitive technical intelligence process, his or her responsibilities can be negotiated. Table 7-1 lists suggested responsibilities. As most of these responsibilities involve interactive roles with business and technical management, they will tend to have strong appeal to the gatekeeper.

State of the Art.   Periodically, the gatekeeper should be expected to provide an overview of the state of the art for the selected technology for which he or she is responsible. Michael Porter's model, depicting the competitive forces within an industry, represents a useful framework for providing the overview. In this manner, the gatekeeper can update the organization on the level of maturity of the technology, significant new developments or changes in the competitive scene, impact of external forces, and the outlook for the future of the technology.

Understanding of Competitive Position.   The gatekeeper is a natural focal point for the integration of competitive information regarding his or her selected technology, and for finding the missing pieces to fill in the gaps. This does not necessarily mean that gatekeepers should personally engage in patent surveys, trend analysis, and technology searches. It's fine if they have a propensity to perform these functions themselves. What is important is that the gatekeepers see to it that the surveys and analyses get done, regardless of who does it, and review and get agreement with the results of the analysis. When information gaps are identified, the gatekeeper usually has a good idea of where to go to fill the gaps, whether it be the literature, or someone in the information network, inside or outside the organization. It is incumbent upon the gatekeeper to interpret the significance of the organization's competitive position, and, if the position is inadequate, recommend actions to change the current position.

Future Selected Technologies.   These technologies, which are the focus of competitive intelligence studies and the elements of the strategic technology plan, will not turn over rapidly in the portfolio if carefully selected. Nor are they permanent fixtures. In a dynamic environment, some technologies may be dropped every year or so, while others are added.

Gatekeepers need to be involved in validating the continued viability of the existing technology portfolio, as well as recommending candidates for deletion or addition. Gatekeepers are often in the best position to provide the holistic view of what is occurring in the external environment, beyond existing competitors. They can scout for the threat of new entrants, emergence of not-in-kind products and processes, competition from suppliers and customers, as well as the influence of the external forces impacting on the organization—regulatory, economic, global, technological, and societal.

Competitive Profiling.   Although it is not necessary for gatekeepers to physically carry out patent searches, they need to not only ensure that monitoring is going on, but also to maintain a dialog with the investigator to help with such issues as key word identification for search protocols, interpretation of difficult patents, and analyses of trends, as well as size and scope of competitive effort. Ultimately, gatekeepers can help draw conclusions and make recommendations to the business and/or to senior R&D management.

Technology Forecasting.   Most planning processes involve some sort of scenario development to identify the most likely future situation, as well as potential boundary conditions. Gatekeepers are often in the best position to forecast future technology trends to assist the organization in determining which technological "arms it will bear" and to help the organization decide how it will respond to various threats from competition or the external environment.

Technology Impact Assessment.   In deciding which technologies will be included in the organization's portfolio, and subsequently tracked through competitive profiling, an impact assessment is useful to establish priorities by emphasizing those technologies of highest importance to the business. Often, the gatekeeper can provide the business with the insights necessary to understand the relevance and impact of technology on the business, resulting in a more informed impact assessment. This assessment is useful in deciding where to focus R&D efforts and how to deploy available resources.

Draw Conclusions and Recommend Strategies.   While seemingly an obvious extension of working on competitive analysis and technology impact assessment, the issue is the kind of reception that gatekeepers receive from the business and senior R&D management, whether or not their recommendations are acted upon. Here is where we get to the main reason gatekeepers are likely to buy into their role in competitive assessment; in the context of formal responsibilities *they are likely to have greater credibility and influence on the decisions made by the business and senior R&D management, compared to their influence in their usual ad hoc roles.*

Whether it's managing a strategic technology plan or a competitive intelligence system, either activity requires periodic meetings. This periodic meeting process enables the gatekeeper to meet with the key decision makers on a planned and regular basis, and provides the individual with a forum to influence these decision makers in a natural setting, and with a frequency that would otherwise be difficult to duplicate. Similarly, it provides management with a convenient mechanism and practical agenda to stay in touch, and guide the overall effort.

Train Future Internal Gatekeepers.   The gatekeepers are usually in the best position to identify their potential successors and, if allowed, will staff their team with the most appropriate people. Because most of the effective gatekeepers tend to be senior technical people, it is important to provide continuity by transferring expertise to younger team members. In this manner, the senior gatekeeper can act as coach and counselor, while the younger members can help get the various tasks done.

Implementing the Program

Establishing a formalized internal gatekeeper network is a major undertaking and often a major departure from past practices. To emphasize the importance of the activity to the gatekeepers, a major event to kick off the program is worthwhile. In our own case, we held a special dinner meeting for the gatekeepers at an upscale location, hosted by our senior technical officer. On this occasion, the formal program was announced, the selected technologies presented, and the role of the internal gatekeepers described. We invited the gatekeepers to consider volunteering to take responsibility for a selected technology, but we made no closure on this issue that evening. The dialog was carried back to the work site, and in the subsequent days each selected technology was matched up to an appropriate gatekeeper, who, in turn selected one or more team members to assist in carrying out the duties.

Ultimately, this program worked because the intelligence gathered was subsequently acted upon, and feedback was provided to the gatekeepers. The activity became a two-way dialog.

As good as the dialog might be with internal gatekeepers, there is always the risk of the organization developing an internal orientation. To avoid this possibility, periodic reality checks should be considered with external experts utilizing external gatekeepers.

GUIDING PRINCIPLES

Principle #14:   Internal gatekeepers are identified, not appointed.

Principle #15:   Internal gatekeepers can be a key source of competitive technical intelligence if they are integrated into the process, and see their role as something more than a paper exercise.

8

External Gatekeepers as a Source of Competitive Technical Intelligence

External Gatekeepers Defined

External gatekeepers are personal contacts who represent secondary sources of information, including competitive technical intelligence. They can be an objective source of information because they are not part of the organization, and are not subject to the same norms and pressures that face employees.

Examples of external gatekeepers include consultants, editors, suppliers/vendors, analysts, and retired executives. In addition, a new external gate-keeper resource has been developed that deserves further description, called key expert panels (KEPs).[1]

KEPs represent a highly cost-effective and productive management tool that can be used to address critical business, technology, and innovation issues through the use of recognized external experts. KEPs extend the proven approach of using groups for problem solving by creating a direct working interface with the best external expert groups in the world. KEPs can provide an objective, external view of technologies to challenge, validate, and augment assessments obtained from internal gatekeepers and patent analyses. Thus, KEPs can be significant triggers for organizational learning and change.

This chapter describes what a KEP is, how it is set up, how it is facilitated, how objectives, deliverables, and agendas are defined, and how long it takes to complete. The power of this unique management tool is demonstrated by studying a specific KEP through the development of its objectives, deliverables, agenda, and outcomes.

What Are Key Expert Panels?

Key expert panels are focused brainstorming sessions involving a panel of world-class experts. The panel usually consists of 8 to 10 internationally known experts recruited from industry, government, and academia. These experts specialize in technologies, and possess experience relevant to the business, products, processes and services. The panel meets for 24 hours to address a specific business issue or set of issues.

Experts possess world-class knowledge that is relevant to the future. They are gatekeepers to the knowledge communities, which are usually a rich network of interpersonal communications and leading-edge researchers and practitioners. They bring a wealth of knowledge of what is going on. Some experts may be leaders whose views do not simply reflect or anticipate events, but who can significantly influence future direction.

KEPs provide a powerful tool in the ongoing process of maintaining organizational alertness, external awareness, and objectivity.

When to Use KEPs

Organizations are faced with increasingly complex business and technology issues that must be effectively and quickly addressed. Time to market and product life cycles are decreasing, and competition has never been as strong on a global basis. To lead the competition, companies must make timely decisions. KEPs afford the opportunity to improve the quality of, and accelerate, the decision-making process.

As shown in Table 8-1, KEPs can be used for a variety of needs beyond competitive technical intelligence.

Within the scope of technology forecasting and assessment, KEPs can be useful in identifying new technologies for competitive profiling. This identification can supplement, build upon, or go beyond technologies identified through internal gatekeepers. In addition, KEPs can validate, modify, or challenge technology impact assessments drawn previously from internal sources.

Table 8-1 Applications for KEPs

Competitive technical intelligence
Technology forecasting and assessment
Business tactical planning
Business strategic planning
Problem solving
Strategic technology planning
Growth initiatives and strategies

Source: Reprinted with permission from reference 1. Copyright 1996 The Fusfeld Group, Inc.

Table 8-2 Examples of Issues Addressed in KEPs

What are the best new business opportunities for the XYZ Business?

What are the future critical issues in the market environment of the late 1990s?

What do successful companies do to create and continuously stimulate an innovative work environment?

What will be the dominant factors in the environmental marketplace in the years 2005 to 2010? How can our company capitalize on the opportunities presented?

Where would you invest R&D dollars today to develop materials and systems that will support the commercialization of four critical technologies in the years beyond 2005: opto-electronics, electronic materials, advanced ceramics, and sensors?

Source: Reprinted with permission from reference 1. Copyright 1996 The Fusfeld Group, Inc.

As exemplified by the issues listed in Table 8-2, KEPs are a cost-effective source of ideas for growth opportunities, outsourcing, partnering, and leveraging core competencies. They can also be used to identify key drivers in developing and commercializing a selected technology.

Benefits of KEPs

KEPs provide first-hand information about developments that are highly relevant to an organization's future and that originate outside the boundaries of its traditional business. This first-hand exposure builds credibility with senior management, who can personally experience the event.

A business issue, concern, or problem can be addressed very quickly using experts outside the normal reaches of your network and your organization's normal sphere of operation. Ideas can be developed in a cost-effective way for solutions to business issues, growth opportunities, outsourcing, and partnering, as well as leveraging core competencies.

By using KEPs, you can obtain the best informed view on which future technologies will have the most impact on your business.

By meeting with, and hearing, the experts first-hand, the organization can follow-up with those sources with whom it would be worthwhile keeping in touch on a continuing basis. Furthermore, by having demonstrated its willingness to listen, the organization can build its reputation among those people who carry influence in environments of interest.

Preparing for a KEP

Getting Started.   An external consultant/facilitator needs to define with the client organization the nature of the issues/problems, the objectives, and the desired outcomes and solutions. The more time that is spent on the definition of the problem and the desired outcomes, the higher the value of the output from the KEP. This definition process involves a series of meetings with the client organization to define the issues, and through an iterative process, the objectives are refined and the deliverables polished.

Depending somewhat on the issue, it typically requires one to two months to set up a KEP, one day for the KEP itself, and a week for the analysis of the output and feedback.

Definition of Objectives.   The definition of objectives is the single-most important factor in determining the success of the KEP, because objectives lead to the definition of the deliverables and the agenda. An example of an objective is shown in Table 8-3.

In this chart, the issue of potential competitors and their core competencies is dealt with in Step 7. Consistent with the logic flow, it is first necessary to identify needs (attributes) and opportunities (product and service concepts), from which supporting technologies can be identified. Only at this point does it make sense to consider potential competitors and their relative strengths and weaknesses.

The use of a KEP in this instance replaced a consulting project that would have cost three to four times that of a KEP, and probably taken two to three times as long to complete.

Deliverables.   The deliverables are derived directly from the objectives. In the example of the environmental marketplace, the deliverables are shown in Table 8-4.

Table 8-3 Objectives for a KEP on the Environmental Marketplace

1. Identify emerging needs in the global environmental marketplace for the time frame 2005–2010. (Consider the needs of industrialized nations only. Categorize outputs according to the following major market segments: solid waste, hazardous waste, petrochemical, chemical, air, and water, with special emphasis on air and water.)
2. Identify the key issues impacting the global environmental marketplace. (The general future direction of standards, regulations, and policy and how they will drive technology should be defined.) What are the likely outcomes and solutions?
3. Identify those environmental markets, or market niches, that appear to provide the best opportunities and best potential fit for the company's businesses and with the corporation's core technology competencies.
4. From the markets identified in 3 above, what products/services provide unique opportunities for the company? How should the company focus its R&D efforts both in its corporate laboratories and in the businesses to pursue these opportunities.
5. Review the outputs from 4 above, identify the "top 10" opportunities and rank order them.
6. Identify the "top 2" or "top 3" opportunities that should be the focus of the second KEP.
7. Start to define the "top 2 or 3" opportunities. Identify the core competencies required. Identify the probable competitors and their existing competencies. What is the potential size of the marketplace? How does the company win?

Source: Reprinted with permission from reference 1. Copyright 1996 The Fusfeld Group, Inc.

Table 8-4 Deliverables for the Environmental Marketplace KEP

---

1. A list of emerging needs for the global environmental marketplace with focus on:
   potable water
   effluent water
   stack gas and plant emissions
   ambient air
2. A list of critical issues (technology, regulations, and policy) that will impact the global environmental marketplace that identifies likely outcomes and potential solutions.
3. A list of market opportunities that provide the best potential fit with the company's core competencies.
4. A "top 10" list of opportunities for the company.
5. Identification of the "top 2" opportunities.
6. Initial definition of the "top 2" opportunities.

Source: Reprinted with permission from reference 1. Copyright 1996 The Fusfeld Group, Inc.

The Agenda.   Once objectives and deliverables have been defined, the task of identifying the specific activities for the KEP is a relatively straightforward process. With start and finish times defined, with appropriate breaks for lunch, etc., the total working time for the experts is fixed. The task, then, is to define how the panel can most effectively meet the objectives and deliver the desired outcomes. A sample agenda is shown as Table 8-5.

Brainstorming is typically one of the initial activities, and there may be a series of separate brainstorming sessions. With disciplined facilitation, a 45-minute brainstorming session can generate from 50 to 100 ideas and/or approaches. These ideas provide much fuel for further discussion and analysis in the later sessions.

Recruiting the Panel Members.   The definition of the objectives, deliverables, and the agenda must be completed before the recruiting phase can begin. A list of prospective key experts is prepared (everybody will want to know who else is coming), the date is set, and the location for the event is chosen. All this information should be presented in an easy-to-read format that is ready to be faxed.

With the package ready, the phone campaign can begin. Our experience is that 30 to 50 percent of those invited will accept, with most objections resulting from scheduling conflicts.

Why do people agree to participate? Panel members from industry typically come from successful companies that are always looking to learn more. They are happy to participate as long as they receive copies of the output. The same applies for consultants and academics who, in addition, will receive an honorarium. The experts represent gatekeepers. They may or may not know the other participants ahead of time, but in addition to gaining new knowledge, they all are eager to strengthen their own networks by renewing previous acquaintances and making new ones as well.

Table 8-5 Agenda for Environmental Marketplace

8:00	Introduction—Review objectives, deliverables, agenda.
8:30	Brainstorm list of needs for global environmental marketplace—focus on **potable water.**
9:15	Brainstorm—focus on **effluent water.**
10:00	Break
10:15	Brainstorm—focus on **stack gas and plant emissions.**
11:00	Brainstorm—focus on **ambient air.**
11:45	Brainstorm lists of critical issues that will impact the global environmental marketplace (technology, regulations, policy).
12:30	Lunch
1:30	Brainstorm a list of market opportunities that provide the best potential fit with the company's core competencies (List competencies).
2:15	Develop a prioritized "top 10' list of opportunities for the company, and identify the "top 2" opportunities.
3:00	Break
3:15	Initial definition of the "top 2" opportunities. Identify the core competencies required. Identify the probable competitors and their existing competencies. What is the potential size of the market place? How does the company win?
4:30	Reflections
5:00	Adjourn

Source: Reprinted with permission from reference 1. Copyright 1996 The Fusfeld Group, Inc.

Security and Proprietary Information Protection.   Frequently, the question of security arises with a client organization the first time it considers the use of a KEP. In reality, security of proprietary information is a non-issue. First, direct competitors of the client firm are excluded from the panel. Occasionally, an early retiree from a competitor may be included.

Second, all panelists are required to sign a confidentiality agreement. More important than the legal implications, signing this document emphasizes the importance of confidentiality to the panelist and provides some reassurance to the client organization.

Most importantly, KEPs are structured so that they focus on future trends, technologies, and competitive activities, all external to the organization. They do not focus on the internal activities of the client company, except to the extent that the client is comfortable in sharing background information. This shared information tends not to be sensitive in nature and is usually a matter of public record in one form or another, such as information revealed in patents, annual reports, product literature, and other publications. Finally, the KEP makes recommendations to the client, but usually is not privy to the decisions and subsequent actions by the client. Therefore, the issue of security is invariably a matter of internal misconception rather than a practical reality.

KEP Facilitation

Effective facilitation requires both focus and discipline. The key is to keep the group focused on the objectives and deliverables while letting the group lead itself, as opposed to the facilitators leading the group. The ultimate success is to have the panel members and company representatives feel good about the end results and that they had fun. The keys to a successful KEP are shown in Table 8-6.

The KEP typically starts at 6:00 P.M. with a dinner and a short presentation on the subject to be addressed at the panel meeting the following day. This provides some "bonding time" for the panel members and also provides a focus for the conversation at dinner.

The panel typically starts at 8:00 A.M. the following morning and ends at 5:00 P.M. to allow panel members to return home the same day. The eight to ten panel members and one or two internal gatekeepers are seated around a U-shaped table. Several key leaders from the company are invited to sit around the perimeter of the room, so that they look into and "eavesdrop" on the experts in the "fishbowl". Facilitators work at the open end of the U using flip charts and other graphic aids.

This seating format provides the stimulation which occurs in a line group discussion. The impact on the organization can be great when its leaders collectively hear without filtering, high-quality information from credible but unfamiliar sources. Nothing can substitute for the emotional charge, the immediacy of the messages that come across, and the opportunity to cross-question the experts.

Output

Some of the most effective contributions come from collateral industries. For example, during one KEP focused on information technology (IT) supported services, a terrific idea came from a freight company executive whose IT infrastructures allowed the company to deliver a level of service which re-defined its industry. His persistent challenging of existing standards in the

Table 8-6 Keys to a Successful KEP

Well-defined objectives and deliverables
A fact-filled recruiting package
Suitable and well-prepared pre-reading materials
Focused and enthusiastic facilitation
An energetic group of experts
Timeliness

Source: Reprinted with permission from reference 1.
Copyright 1996 The Fusfeld Group, Inc.

building services business precipitated a whole new set of ambitions among the other participants.

Typically, a 30–35-page report is published summarizing the data generated during the day. The report is formatted to respond to both the objectives and the deliverables. It is published as soon as possible following the completion of the KEP to capture the dynamics of the day. Invariably, the results have high impact on the client organization, as the KEP usually catalyzes new strategic directions for a business.

GUIDING PRINCIPLES

Principle # 16:   Key expert panels are a cost-and time-effective way to access external gatekeepers for technology forecasting and as a source of competitive technical intelligence.

9

Lead Users and Strategic Suppliers as Competitive Intelligence Sources

Lead Users

Like internal gatekeepers, lead users represent a personal and primary source of information. Similar to external gatekeepers, lead users represent an external viewpoint with the added advantage of their representing the voice of the customer, and potentially your customer at that! Competitive intelligence information is therefore a secondary benefit derived from the relationship with a lead user.

In every industry, there are usually one or more companies who are on the leading edge of technology, and are the first to adopt new technologies. The concept here is to align your organization with such companies in order to obtain a window on their view of future technology needs, not only for themselves but for the industry in general. With this alignment will also come a perspective on the positioning of other competitors, both yours and those of your lead user.

In recent years, more and more companies have come to recognize that "partnering" is an excellent way to gain a customer. The intelligence benefits are a secondary fallout from the relationship. This relationship has been given a variety of names, such as *customer partnering* or *customer intimate discipline*.

Identifying and Engaging Lead Users

Identifying and engaging the lead users in an industry may be more challenging than you imagine. Lead users known by reputation generally already have well-established development partners, and it may be difficult to "break in" if you are not the first to develop an initiative in a particular field.

A colleague in the medical instrument field points out that identifying and working with lead users among physicians is absolutely critical if the company's instrument is to be subsequently adopted by the medical community.

Some lead users may be less obvious. An excellent way to identify them is by utilizing your competitive intelligence resources, such as patent surveys, patent technology profiles, and internal and external gatekeepers.

This approach is further strengthened if it can be linked to an organization's strategic technology plan. In this case, the organization determines the technological orientation of the lead user (attributes), translates these attributes into technology needs utilizing the customer needs and technology matrix, and aligns its own strategy with that of the lead user. When a potential fit is developed, the organization is in a position to approach the lead user, and present a plan that shows how its needs can be met through a closer working relationship, such as a joint development program, with the potential supplier.

There is always a risk of losing proprietary information in these engagements, because in order to identify a partner who provides a good strategic fit, it's necessary to share information on your own strategic direction. Confidentiality agreements to cover the information exchange can usually be negotiated. A more difficult problem arises when the lead user requests exclusive access to the resulting new products and underlying technology. To counter this request, the lead user can be reassured of sufficient lead time as the reward for close collaboration, rather than an exclusive position.

Getting Started—Keep It Simple!

Once you have identified a potential partner, you cannot rush the necessary interactions, but must keep the approach simple. Be sure that any cooperation satisfies the strategic intents of both parties. With a lead user, the most common arrangement is that the lead user will get some agreed-upon period of lead time from the results of the joint program, while the supplier gets both lead time and a proprietary position on the product or process. In niche markets, the lead user may get an exclusive position as well.

In most cases, the initial cooperation should be limited to one program, with both parties having clearly identified key contacts to manage the dialog. With simplicity in mind, it's best to start with a relatively few participants from each organization and build up as progress is made. The sponsors need to periodically assess the value of continuing the relationship.

Keeping it simple also applies to the number of lead users that you engage. Often, one lead user relationship within an industry segment is sufficient, although this may not be the case in the medical instrument field, where a number of lead users may need to be engaged, if they exist. Conversely, depending upon the variety of materials and services purchased by an enterprise, a lead user could have a number of strategic suppliers, as long as no conflict of interest exists between the fields of endeavor among the suppliers.

Good Customers versus Lead Users

A good customer is not necessarily a lead user. Sometimes, in a breakdown in communications between the R&D and business divisions, the business division may choose to work closely with a good customer who is not a lead user because the business division fears incurring the customer's wrath if the business chooses to work with another company.

In some cases this choice to work with a good customer rather than a lead user can result in commercial disaster. A classic example involved a supplier of coatings for the interior of cans used in the beer and beverage industry. A number of years ago, the primary manufacturers of such cans were American Can, Continental Can, and National Can Company. These companies represented the major customers for the supplier of can coatings. The cans were made from steel.

A major innovation was occurring in this industry with the development and commercialization of aluminum cans. Unfortunately, the major can producers were not involved in this development. The aluminum companies worked directly with the beer and beverage producers on this development. With this information, a coatings company, which was a competitor to the major can-coating supplier, correctly identified the lead users and collaborated with them to develop appropriate, environmentally friendly coatings for the new aluminum beverage cans. The program was highly successful, resulting in a new market for the aluminum companies, and establishing the new coatings manufacturer as the preeminent supplier to this industry. The original can companies lost out in what had been a major market for them, and their key coating supplier's business dwindled away to the point that the supplier went out of business. Throughout this time period, the coating supplier was well aware of the activities of the aluminum companies and the beer and beverage producers, yet doggedly held onto its relationship with the can companies. By the time the coating supplier tried to address the aluminum can market, it was too late.

In another case, a supplier of coatings acquired new environmentally friendly coatings technology in the early stages of development, with the goal of developing a competitive advantage by being first to market with the environmentally friendly technology. The business targeted its major customer, with whom it had great rapport, but who, unfortunately, was not reputed to be a lead user. This customer worked closely with the supplier, but continued to delay commercialization with one excuse after another. By the time the new technology was commercialized with this customer, competition had caught up, so that no differentiating position could be established. In this case, the coatings company had squandered three years of lead time.

Strategic Suppliers as a Source of Competitive Intelligence

The concept of a strategic supplier is not new. Industrial customers have been collaborating with selected suppliers for many years, but a strategic supplier is different. Just as with lead users may be different from good customers, strategic suppliers may be different from a good supplier that has had a good track record with regard to price, quality, and service. Competitive intelligence, as with lead users, is a secondary fallout from the relationship.

Often, companies tend to be reactive to the initiatives of suppliers. Many vendors are intent upon pushing their newest developments on customers, whether or not these developments fit with the customers' needs. A more productive approach is for the organization to take the initiative to identify and approach potentially strategic suppliers.

A strategic supplier's strategic intents and technology competencies fit well with the needs of the lead user. Utilizing competitive technical intelligence resources, technology leaders among the various suppliers can be identified, as well as those vendors whose core competencies complement the customer's needs. Armed with this intelligence, a company is in a position to approach selected suppliers and solicit their cooperation in new joint development initiatives.

Gaining a commitment from a supplier to work exclusively with your company may not be an easy task. Two approaches have proven successful, and, when used in combination, can be powerful tools. The first approach involves sharing the organization's strategic plan, under an appropriate confidentiality agreement, so that the supplier can understand the strategic fit between the organizations and the opportunity for a positive business impact, should the joint effort succeed. The second approach involves, immediately rewarding the supplier for engaging in such a cooperation, rather than expecting the supplier to gamble on a positive outcome. This means shifting some existing business to the vendor, if it is new to the company, or increasing current business participation if the supplier relationship already exists. In this manner, the criterion of technical support is added to the traditional criteria for vendor selection of price, quality, and service. These approaches do not ensure success. In one case, a key supplier staunchly refused to participate in such a program, not wishing to show favoritism to one customer at the expense of others.

In another example, a chemical manufacturer divided its purchases of chemical intermediates between two major suppliers. The chemical manufacturer enjoyed a long-standing and amicable technical cooperation with one of these suppliers for many years, which had resulted in some modest developments. To the manufacturer's surprise, the second supplier introduced a product representing a major advance in a chemical intermediate, that could not be supplied by the cooperating supplier. A subsequent analysis of the technology base of the two suppliers showed that the manufacturer was working with the wrong supplier. The cooperating supplier was primarily harvesting the manufacturer's technology, while the other supplier was investing heavily in new technology, as shown by the quantity and type of patents that its was issuing. Subsequently, amid many complaints and cries of anguish from its primary supplier, the manufacturer shifted its attention and subsequent development efforts to the second supplier, along with a reallocation of business awards between the two suppliers.

In another case, a paint company, known to be a leader in new color styling and color effects for the automotive industry, approached a specific pigment manufacturer with a proposal to collaborate on new color developments. The paint company would be the first to introduce the newly styled colors and gain lead time on its competitors, while the pigment company could shorten its time to commercialization for new pigments (then running about 10 years), and have a proprietary position as well.

The paint company had identified a need for a new shade of red for new models of automobiles to meet new styling trends, which could not be met with commercially available pigments. The pigment supplier found an existing dye on their "shelf" that met the color space requirements for the new chromophore, and set about developing a pigment based on this dyestuff.

The program was a great success. The new red pigment was developed and commercialized in 4 years, compared with the standard 10-year time frame. The paint company, awarded a one-year lead time, captured the market for this color for the first model year, and having established this position, remained the premier supplier for this color for a number of years thereafter. The pigment supplier maintained an exclusive, proprietary position to make and sell this particular pigment. As a result of this positive experience, the joint development relationship between these two companies has grown, and continues to endure over the years.

As with lead users, the number of strategic suppliers should be limited based on the types and variety of materials purchased, and the capability of the customer to facilitate joint efforts. In some businesses, one such relationship might suffice.

Protection of Proprietary Information

In dealing with lead users or strategic suppliers, the issue of confidentiality needs to be addressed. Documentation through a confidentiality agreement is a must. Beyond this documentation there has to be some element of trust and understanding of the integrity of the other party. If integrity is in question, then you have the wrong partner.

On the one hand, suppliers eager to curry favor with a potential or good customer may reveal information regarding their customer's competitors, some of which may be confidential in nature. While such suppliers may be an excellent source of competitive intelligence, they would not qualify as strategic suppliers. If this behavior is evident, you can assume that the supplier will subsequently share information on your own organization with competitors as well. At the other extreme are suppliers who might represent an excellent strategic fit, but who staunchly refuse to sign a confidentiality agreement. In both cases, such suppliers need to be dealt with on an "arm's-length" basis.

The same concern applies to lead users, customers, or potential customers. The same guidelines apply here as with strategic suppliers. Some customers may try to use confidential information to play one supplier against another. Therefore, you need to consider the customer's reputation, in seeking a strategic alliance, in addition to securing an appropriate confidentiality agreement.

GUIDING PRINCIPLES

Principle #17: Lead users and strategic suppliers offer an excellent opportunity to advance new technology for competitive advantage through joint programs.

Principle #18: Lead users need to be identified and differentiated from good customers.

Principle #19: Find ways to work with lead users, without losing good customers.

Principle #20: Strategic suppliers need to be identified on the basis of their strategic intent and competitive technology position, in addition to the traditional criteria of price, quality, and service.

Principle #21: Strategic suppliers need to be provided with both strategic and near-term incentives to engage in exclusive joint programs.

Organizing the Competitive
Intelligence Activity

Senior Management Support

Regardless of how well the competitive technical intelligence (CTI) program is organized and the various sources of intelligence are utilized, the activity will be relegated to data collection and analysis, that is, "trivial pursuit," unless there is support from senior management! Even if it is a formalized program, without this support the program may as well be ad hoc, with the possibility of it being a management resource only in a reactive mode.

Every senior executive has his or her own network from which to draw actionable information. As indicated earlier, the senior executive will utilize information sources that he or she has confidence in, believes to be credible, and fits with his or her style of receiving information. The CTI activity needs to become part of that network.

There are three ways that support for the CTI activity can be initiated, if it doesn't already exist. First a senior executive takes initiative. Somewhere along in their careers, some senior executives have come to value formal CTI systems, and will proactively promote and support such activities. These executives tend to be forward-looking visionaries. CEO's with this vision tend to surround themselves with business and functional officers who also value and promote competitive intelligence. If the value does not yet exist among his or her subordinates, the CEO will take steps to promote awareness of these external intelligence needs, even going to the extent of designing programs to involve the team in related activities, such as business scenario and technology forecasting.

A second possibility is that senior management takes action after getting a "wake-up call". This alert may manifest itself as being blindsided or upon learning about some surprising information which points out a potential threat to the business that was not anticipated.

Examples of being blindsided were cited earlier. It could be a close call, such as the case of being offered a license by a foreign company for a technology that fits into one's core competency, but one of which the enterprise was previously unaware. Alternatively, it could be the surprise of a public announcement of two competitors forming an alliance, whether or not the enterprise could have taken action had it known about the alliance earlier. In a worst case scenario, being blindsided could take the form of being surprised by the commercial launch of a new, superior product by an existing or new competitor.

The third approach is when the CTI function takes the initiative. In this case, it is imperative that this initiative be carefully designed for success. A surefire way to fail is to try to institute one CTI system across the board, imposing it on senior management. The "one size fits all" approach simply doesn't work, because key executives will reject this approach if it is at variance with the way they usually gather and analyze information.

Surveys indicate that it is most effective for the CTI function to tailor a flexible system to address the varied and individual needs of each of the senior executives within the organization. A step-by-step approach works best. It starts with one-on-one interviews with the senior executives to inform them of the capabilities of the CTI function and to identify their needs and interests. The next step consists of initiating one or more modest programs to address these needs, and following up monthly with the executives to ensure that these initiatives are on target. After three months, the follow-up can be stretched out to quarterly updates, then semi-annually, and finally annual reviews. As success is demonstrated, additional programs may be added, as well as more sophisticated tasks.

Initially, the number of programs initiated with each executive will depend on the CTI resources available, and the degree to which the programs fit or don't fit in with existing activities and competencies. To ensure success, simplicity is critical, as programs are initiated. At the start, it is often desirable to develop a "quick hit" to build rapport, credibility, and value for the activity. While this one-on-one approach may appear to be slow and time-consuming, it has a successful track record. After all, like Rome, a successful CTI program will not be built in a day. An ineffective program, hastily put together, will not only fail, but is likely to shut the door on any chance to try again.

Guerrilla Tactics

What can be done when senior management support does not exist? In these situations, usually there is also no funding or staff allocation for competitive technical intelligence. In such situations, guerrilla tactics are in order!

You can still start a modest effort, utilizing whatever sources are at your disposal. With this approach, it is critical to target for an early success. If the first attempt does not succeed, then you can continue to try, utilizing bootleg, hit-and-run tactics, similar to guerrilla warfare. People responsible for some aspect of technical information services are in the best position to engage in guerrilla activities.

One successful operation started with patenting activities, in which competitor patent profiling was developed on a part-time basis. When these data revealed that the company was actually trailing its competitors to a greater extent than anticipated, the information was presented to senior management. The information acted as a "wake-up call". Subsequently, additional funding and support was provided.

In another case, such an activity developed within the technical library. When internal clients realized what the library was capable of providing, requests started to increase, and subsequently these requests became the justification for increased funding and personnel.

CTI on a "Shoestring"

Surveys have shown that the CTI activity is not necessarily located in R&D, but often located in the marketing organization, although staffed by technical people. Regardless of organizational location, it is imperative that the CTI dialog be carried on across functions, including R&D, marketing, sales, manufacturing, and finance. A viable competitive technical intelligence effort can be initiated at a modest cost. A popular (and accurate) expression is that minimal staffing can be "two people and a library". The two roles that need to be fulfilled are those of a *CTI coordinator* and an *information scientist*.

CTI Coordinator. The role of the CTI coordinator consists of the elements outlined in Table 10-1. One of this person's tasks is to initiate and maintain the dialog with key executives and the appropriate people within the organization. As indicated previously, it falls on the coordinator's shoulders to interview senior managers and their direct reports on a one-on-one basis to define their specific interests and needs regarding competitive intelligence and how best to communicate with them as programs are initiated.

Table 10-1 Duties of the CTI Coordinator

1. Initiate and maintain dialog with senior management
2. Design the CTI system
3. Help prioritize tasks
4. Manage the sources of CTI
5. Help analyze data gathered—put the puzzle together
6. Provide conclusions and recommendations
7. Be the cheerleader!

It is through this interview process that the coordinator is able to design the programs and methods for communicating results of intelligence efforts. Subsequent follow-up interviews with the "clients" are also the coordinator's responsibility.

Working with the information scientist, the coordinator helps prioritize the tasks arising from the executive interviews. While the information scientist conducts the patent surveys, technology profiles, and technology searches, the CTI coordinator orchestrates the activities of the other CTI sources and integrates them with the work of the information scientist. Orchestration of internal gatekeepers involves coordinating their input to help direct the information scientist's efforts, fill in gaps in the intelligence pattern, and help validate or refute the results by having them closely involved in helping to interpret the results. In effect, the coordinator becomes the hub in the communication network with the internal gatekeepers.

Working with outside consultants, the coordinator is responsible for scheduling external gatekeeper sessions as may be appropriate.

While lead user interaction is usually a marketing-directed function, along with R&D participation, the CTI coordinator needs to be linked to both marketing and R&D to pinpoint questions that may be asked of the lead user, as well as to receive input from the lead user regarding his or her contribution to competitive technical intelligence or technology projections.

Of all the people involved in the CTI activity, the coordinator is in the best position to integrate the results of the effort into a coherent picture, just as one puts together the pieces of a jigsaw puzzle. While various individuals may contribute by providing information and by participating in the analysis, it is most effective when one person is ultimately responsible for putting it all together. Therefore, it is only natural that the coordinator take the lead in drawing conclusions, and, to the extent it may be appropriate, providing recommendations to the sponsoring organization or executive.

Vision, initiative, leadership, analysis, integration, and communication mark the characteristics of the successful CTI coordinator. In order to be effective, the coordinator needs to have a well-rounded background. A technical background is important, because it is easier to train a technical person in competitive intelligence activities, than to do the reverse. This individual needs to have an understanding of the business being supported, as well as the technologies important to the business. This is clearly not the assignment for a new employee.

Information Scientist.   The duties of the information scientist are outlined in Table 10-2. Given these duties, the qualifications for an information scientist are quite different, yet complementary to those for the CTI coordinator. The information scientist is a technically trained individual who knows how to conduct patent and literature searches and is knowledgeable in computer-driven searching. Because the individual is required to continuously make instant assessments of the relevance of patent disclosures and the impact of other technical information on the business, the information scientist should

Table 10-2 Duties of the Information Scientist

1. Designs and carries out patent searches
2. Organizes and communicates the results of patent surveys, technology profiles, and technology searches
3. Conducts special technology searches, utilizing current literature and sources available on the internet
4. Designs, develops, and distributes selective dissemination of information patent studies (SDI's).

have a strong background in the relevant technology and a good understanding of the business being supported. In the chemical industry, it is preferable that this individual have an advanced degree in chemistry or chemical engineering, as well as experience in applications R&D. This enables the information scientist to subsequently interpret the literature and patents in terms of their potential utility and applications.

The question is often raised as to whether one starts with a technically trained individual or someone trained in library science. Clearly, the technically trained person is the better choice, because one can more readily learn on the job to search the literature and to acquire computer skills than to learn chemistry or chemical engineering.

However, most technically trained people do not make good information scientists! The issue is not skills or knowledge, but rather how one likes to spend one's workday. The ideal information scientist enjoys working alone, with information rather than people. He or she is an intuitive person, interested in concepts and theories. Information scientists are logical people with a great deal of patience and perseverance, who enjoy working in an unstructured environment and who are driven by their own curiosity. Most technical people are also interested in theories and concepts, but do not enjoy the continuous amount of reading required in literature searching. Generally, they do not have the patience required to successfully complete a thorough search. Therefore, the prospective information scientist needs to be carefully selected. Hang on to people who are energized by this type of work—they are exceptional!

Routine versus Special Studies.   Many companies already have some sort of routine patent surveillance in place. Information scientists usually generate a number of different SDI patent abstracts on topics of interest to members of the technical community on a periodic basis. While this is a worthwhile activity, it is not sufficient.

Patents received by an R&D person on a regular basis tend to be read like a newspaper. If something particular strikes the reader, some sort of action may be taken. What is not getting done is the systematic grouping, totaling up, and trend plotting which are the more productive types of analyses.

In addition to the routine SDIs, an important function for the CTI coordinator, in collaboration with the information scientist, is to develop and

Table 10-3 The Complete CI Book

1. Key business strategies of competitors
2. Competitive technology assessment
3. Competitive manufacturing profiles
4. Competitive marketing profiles.
5. Comparative financial analyses

assemble the complete competitive intelligence (CI) book on each key competitor, providing a technology assessment for each, as shown in Table 10-3. Equally important is to keep these books up to date on some periodic basis. Beyond this documentation, special studies will depend upon what other activities are driving the competitive intelligence effort, such as a strategic technology planning process, or a centralized, coordinated competitive intelligence program.

Inevitably, particularly in a "shoestring" operation, the demands for the routine services of the information scientist can dominate to the point that the special studies, which are important, but less urgent, fail to get done. It is incumbent upon both the CTI coordinator and information scientist to maintain frequent communications to ensure a balance in priorities is maintained.

More Sophisticated Competitive Intelligence Systems

There are a variety of competitive intelligence systems, depending upon the needs and desires of key executives of the organization. So far, the shoestring-type operation has been described. Three additional examples should provide the reader with an idea what might be possible and appropriate.

Adjunct to Strategic Planning.   In this scenario, the driver is the strategic technology plan and the strategic technology planning process. The CTI program can still be operating on a shoestring as described earlier. The strategic technology plan is usually sponsored and controlled by a technology council, consisting of a cross-functional team of senior executives. The CTI coordinator reports to a council member, usually in the technology function, because the strategic technology planning process depends heavily on competitive technical intelligence to not only determine the organization's competitive position in selected technologies, but to gain insights into possible actions to be taken regarding deployment of technical assets for competitive advantage.

Centralized CI Function.   Ideally, a centralized CI function would be located within the business. In this case, the coordinator would have a staff capable of managing the complete activity, although the coordinator would call upon resources within the functions for specific contributions. Under the aegis of the central coordinator, the complete CI book can be put together, as shown in Table 10-3.

Recall the credibility issue described earlier, wherein the senior management of a manufacturer of automotive engineering plastic components was shocked into disbelief to learn that, regarding technology, the company was further behind its competition than it had believed. After receiving confirmation from an outside consulting firm, the company became proactive. It established a central competitive intelligence function, and subsequently gained equally new and startling insights into the relative competitiveness of its various manufacturing facilities, differences as to how R&D was being funded by competitors, and differences in marketing strategies as well. By the time it completed its analyses, the company had de facto developed the complete CI Book.

Once the company took ownership of the intelligence developed, it was able to initiate a series of actions to strengthen its competitive position. The company reduced manufacturing costs, developed greater focus on the market segments of concern to the business, increased market share within these segments, and redeployed R&D resources to improve the focus and subsequent impact of their efforts. Today, the company is in a substantially stronger competitive position, as measured by revenues, market share, and profitability.

Decentralized Competitive Intelligence Function. In a decentralized activity, the focal point remains with a central coordinator located within the business itself, reporting to a senior management cross-functional council. In contrast to the centralized configuration, the coordinator has no staff in the decentralized organization. Responsibilities reside within the functional units: R&D, manufacturing, marketing, sales, and finance. One application for this configuration could be the commodity chemical business. To be profitable in a chemical commodity, it is important to be the low-cost producer. Leading companies have sophisticated CI systems in place to ensure that they are, and remain, the low-cost producer in their business.

In a commodity chemical business with this type of CI configuration, manufacturing takes the lead in driving to be the low-cost producer. Competitive intelligence groups are formed at each manufacturing site. In turn, each site adopts a specific competitor's site to study and compare cost and quality of manufacture. By coordinating this activity across the function, all competitors' manufacturing sites are covered within the company. Information regarding these sites can subsequently be rearranged along product and process lines, so that current alternative manufacturing processes are under ongoing assessment for cost-effectiveness and quality as well.

In one particular case, a company with a strong CI system in place was faced with the task of building a new plant to increase its capacity in a commodity chemical. Based upon their studies, the technical, engineering, and manufacturing groups reached a consensus that the company needed to adopt a competitor's process for the new facility. This process would result in lower cost and better quality than its current process technology!

Outsourcing the CTI Function?

In these times of downsizing, it is reasonable to ask whether the CTI function can be cost-effectively handled by outsourcing. Clearly, specific studies can be outsourced. On the other hand, the knowledge base required to integrate information into a comprehensive picture, and the need for ongoing monitoring of the technical literature, should preclude outsourcing the coordinator and information scientist functions.

To ensure that these functions become institutionalized and not downsizing targets, it is important to attach them to ongoing activities such as library operations, marketing and market research, strategic planning, R&D, or manufacturing. As a stand-alone function, competitive technical intelligence will endure only to the extent that senior management believes in and utilizes it. As with many R&D functions, CTI may also help to maintain awareness by periodically reporting benefits, and utilizing metrics to quantify activities, as described in Chapter 2.

GUIDING PRINCIPLES

Principle #22: The easiest way to institutionalize competitive technical intelligence is to attach it to an existing planning process.

Principle #23: Regardless of the organizational configuration, ultimately one person needs to have all the information available so that it can be integrated into a consistent story.

The Role of Competitive Technical Intelligence in the Strategic Planning Process

Competitive Technical Intelligence as a Proactive Tool

The point has already been made that competitive technical intelligence must be actionable or it is nothing more than a trivial pursuit. It was also shown that most companies that have an intelligence program utilize it primarily in a reactive mode. In more visionary organizations, competitive technical intelligence can be a proactive tool when it is integrated into a strategic planning process. In order to better understand the role of intelligence, you must first have an understanding of the strategic planning process.

Linking Customer Needs with Technology

Until now, the issue of addressing customer needs has been implied but not squarely addressed. Strategic technology planning starts with the customer. Customer needs or attributes can be, in turn, linked to product, process, or service concepts as shown in Figure 11-1. Subsequently, these concepts can be linked back to the technologies necessary to deliver and support the concepts. In this way, a direct connection is made between technologies and customer attributes. Note that in each of these grids, impact and competitive position are the key considerations. First let's make the connection by considering the subject of impact.

Customer Impact.   The linkage regarding impact of customer needs with that of products, processes, and services, and ultimately technologies, is de-

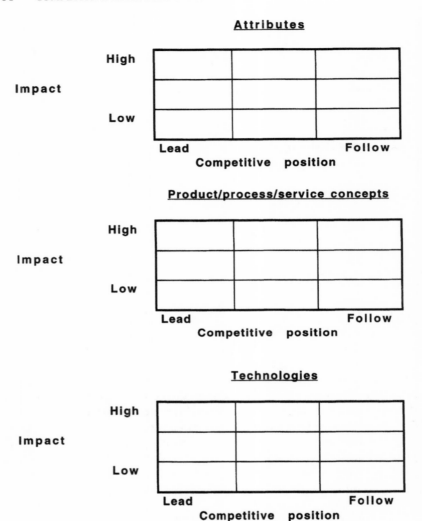

Figure 11-1. Business linkage.

veloped by first using the customer/technology matrix shown in Figure 11-2. This framework is similar to methodologies described as the House of Quality[1], the Voice of the Customer[2], or Quality Function Deployment. Customer needs or attributes are listed under the topic heading and given a weighting of high, medium, or low depending upon the importance of that attribute to the customer. Product, process, or service concepts are then generated and inserted into the matrix. A rating system is then used to determine how well a given concept meets the specific attribute, in this case, 3-2-1 equates to high-medium-low. In this manner, a weighting for each attribute can be developed.

A second matrix can subsequently be developed in which the customer attributes for each concept are rated against supporting technologies. Ultimately, this matrix enables the analyst to determine which concepts and technologies are likely to have the highest customer impact.

To illustrate how this matrix works, let's assume that a strategic technology plan is being developed for a fictional company, Automotive Aftermarket

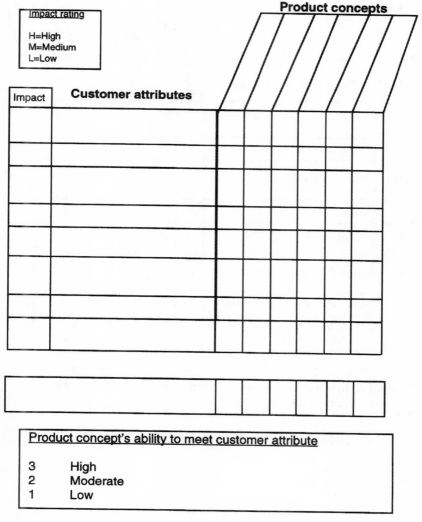

Figure 11-2. Relevance of product concepts to customer attributes.

Paints, Inc. (AAP), which produces and sells paints for the automotive repair and refinishing market. For simplicity, only a portion of the customer attributes, product concepts, and technologies will be considered, as shown in Figure 11-3.

First, the customer attributes (needs and wants) are defined and weighted in terms of importance (impact) to the customer. In this example, the cus-

Product concepts

Impact	Customer attributes	Water based enamel	Acrylic enamel	Acrylic lacquer	Polyester enamel	Nitrocellulose lacquer	Two-component enamel
M	"Once around the car"	1	2	1	3	1	2
M	Application latitude	1	2	1	3	1	2
HH	Speed of dry	2	2	3	2	3	1
H	Color match	2	3	2	2	2	3
H	Initial gloss (no buffing)	2	2	1	3	1	3
H	Color & gloss retention	2	2	2	1	1	3
H	Acid rain etch resistance	1	1	1	1	1	2
H	Resistance to chipping, flaking and peeling	3	3	2	3	2	3
Raw weighting		42	49	40	47	37	54
Impact of product concepts (Percent weighting; total=100%)		15.6	18.2	14.9	17.4	13.8	20.1

Impact rating

H=High
M=Medium
L=Low

Product concept's ability to meet customer attribute

3   High
2   Moderate
1   Low

Figure 11-3. Relevance of product concepts to customer attributes. Premium car repainting system.

tomers are defined as the automobile-body repair shop owners in North America, and the segment being analyzed is the premium topcoat segment. While the importance of attributes are defined in terms of high, medium, and low, the premium market segment has been plagued by slow-drying paints as the tradeoff for premium quality. As a result, the customer has placed an unusually high premium on speed of dry which receives a high high (HH) rating. As a result, this attribute receives a weighting of 4 on the 3-2-1 scale.

Next, the product concepts that represent candidates to meet the need for an improved premium topcoat are developed and listed across the top of the matrix. The ability of each product concept to deliver each attribute is then rated on a 3-2-1 scale, with 3 being high and 1 being low.

A raw weighting score is then developed for each product concept by multiplying the rating for the suitability of the product concept by the rating of the impact of the customer attribute and totaling the score for each product concept. In this case, the impact weighting of high-medium-low is also converted to a 3-2-1 numerical scale. The raw scores are then converted into a percentage.

In this example, the highest score is accorded to a two-component system as the product concept having the highest probability of meeting the customer attributes, with acrylic enamels rated second and polyester enamels rated third. The results of this scoring pose a dilemma. Of all the product concepts, the only one that provides a reasonable level of acid rain etch-resistance is a two-component system, yet it is rated the slowest in speed of dry, the most important customer attribute on the list! Unless the speed of dry of a two-component system can be improved, it would have to be dropped as a candidate.

Acid rain etch-resistance is a relatively new customer need. Acid rain etches the surface of the paint, resulting in dull spots on what is otherwise a glossy surface, and requiring that the car be repainted to repair the damage. Speed of dry is important because many body shops in North America are a one-spray booth operation with minimal baking facilities. The repainted car cannot be moved out of the booth, into the parking lot, until the paint is sufficiently dry so that it will not collect dirt as it is exposed to the atmosphere. The speed of dry in these small shops is a critical issue impacting throughput, the number of cars that can be painted in a day. AAP, Inc. is striving to analyze its options to address this balance of properties as part of a strategic technology plan.

Technology Impact.   Because a two-component system seems to be the best product concept to provide some protection from acid rain etch, AAP has decided to evaluate two-component paint technologies that might also provide the necessary speed of dry. The linkage between customer attributes, product concepts, and technologies is depicted in Figure 11-4. A list of supporting technologies that are relevant to two-component systems is first generated, then tabulated in the technology boxes.

Technologies

Impact rating
H=High
M=Medium
L=Low

Impact	Customer attributes	Cross linking	New monomers	Metered spray equipment	Low-MW components	Colorimetry	Pigmentation
M	"Once around the car"	1	2	3	3	1	2
M	Application latitude	1	3	3	3	1	2
HH	Speed of dry	3	2	1	2	1	2
H	Color match	-1	-1	2	-1	3	3
H	Initial gloss (no buffing)	2	2	1	2	1	2
H	Color & gloss retention	3	3	1	3	1	3
H	Acid rain etch resistance	3	2	1	2	1	-1
H	Resistance to chipping, flaking and peeling	3	3	1	3	1	1
Raw weighting		46	45	34	47	29	40
Impact of product concepts (Percent weighting; total=100%)		19.1	18.7	14.1	19.5	12.0	16.6

Product concept's ability to meet customer attribute
3   High
2   Moderate
1   Low

Negative numbers indicate adverse impact

Figure 11-4. Technology impact on customer attributes for two-component paint systems. Premium car repainting system.

Suitability assessments are developed just as they were for the product concepts. In several cases, a technology may not favorably impact an attribute, but may have the potential for an adverse impact. These cases are noted by a negative number and are subtracted from rather than added to the weighting. For example, a cross-linker may not help produce a color match, but if the cross-linker is highly colored or prone to yellowing, it could adversely affect the color match.

The impact chart shows that the desired balance of properties will be most influenced by low-molecular-weight components, cross-linking agents, new monomers, and pigmentation, while metered spray equipment is not particularly relevant, nor is colorimetry.

Competitive Position.   Additional insight can be obtained for these technologies by looking at their customer impact in relation to AAP's competitive position in these technologies, as shown in Figure 11-5.

When the variable of competitive position in the technology is added to the matrix, AAP, Inc. leads in low-molecular-weight component technology, is equal regarding new monomers, but follows in cross-linkers. It would seem reasonable that the most improvement might come from investigating recent advances in cross-linkers, a technology of least familiarity to AAP, Inc. Given its strong competitive position, AAP, Inc. would be expected to be utilizing leading-edge low-molecular-weight component technology, and is probably also familiar with the latest monomer technology. Because of AAP's following position, any leads regarding cross-linker technology will most likely come from external sources, although, for the sake of thoroughness, external sources might also be investigated for low-molecular-weight components and new monomers as well.

By utilizing the customer/technology matrix, and having knowledge of the customers' needs and one's competitive position with regard to selected technologies, the analyst could quickly reach a tentative conclusion that an

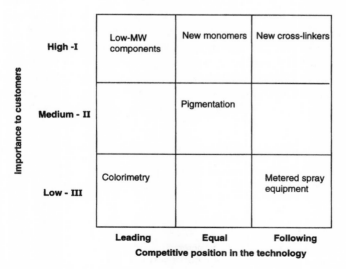

Figure 11-5. Importance to customers versus technology competitive position. Two component automotive repair topcoat enamels.

external technology search regarding cross-linkers may be the most cost-effective approach to upgrading the performance of the preferred product concept, two-component systems.

Additional Strategic Considerations

Before plunging into a technology search, it is worthwhile to consider other strategic technology issues beyond customer impact and competitive position. A more comprehensive list of considerations is shown in Table 11-1. Not all of these considerations may be equally relevant to the issue under analysis, but the relevant ones can only be determined by completing the strategic analysis.

Business Importance and Timing.   The issues of importance to the business and timing can be depicted in Figure 11-6. The strategic time window is defined as the time it takes to go from a clean-sheet design to commercial launch. In the coatings, chemical, and allied industries, this strategic time window is generally 5 to 10 years. Accepting this as the strategic time frame, the tactical window might be specified as 2 to 5 years. Note, again, the relevance of competitive position to timing.

In order to identify which technologies belong in the strategic and tactical windows, scenarios can be developed based on the relevant forces driving the industry over the two time frames. Utilizing our coatings example, the matrix might look something like that depicted in Figure 11-7. As shown, all the technologies cited are important to the business in both the tactical and strategic time frames, and no differentiation in importance or timing is made. The competitive position versus timing matrix does, however, produce some additional pieces of important information. The need to upgrade AAP's competitive position in metered spray equipment and cross-linker technologies is imperative if competitive position is to be maintained, let alone improved! The urgency to become more competitive in the 2 to 5-year time frame strongly supports acquisition of these technologies as opposed to internal development. At this point, the search protocol needs to include

Table 11-1 Strategic Technology Considerations

Impact on customer needs
Impact on business needs
Competitive position
Timing
Maturity of technology
Current level of competency
Desired level of competency
Desired future competitive position
Degree of technical risk
Degree of commercial risk

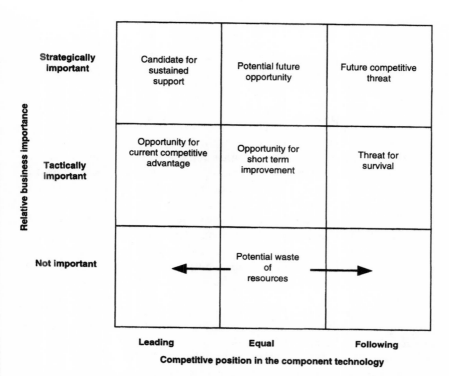

	Leading	Equal	Following
Strategically important	Candidate for sustained support	Potential future opportunity	Future competitive threat
Tactically important	Opportunity for current competitive advantage	Opportunity for short term improvement	Threat for survival
Not important	←	Potential waste of resources	→

Relative business importance (vertical axis)

Competitive position in the component technology

Figure 11-6. Relative business importance versus current competitive position.

	Leading	Equal	Following
Strategically important 5+ years	Colorimetry Low-MW components	Pigmentation New monomers	Metered spray equipment New cross-linkers
Tactically important 2 - 5 years	Colorimetry Low-MW components	Pigmentation New monomers	Metered spray equipment New cross-linkers
Not important			

Relative business importance (vertical axis)

Competitive position in the component technology

Figure 11-7. Relative business importance versus current competitive position. Two-component automotive repair topcoat enamels.

metered spray equipment, as well as cross-linkers. Again, before initiating a search, the remaining strategic technology considerations should be analyzed.

Technology Maturity.   Another consideration is the maturity of the technologies, as shown in Figure 11-8. When the maturity of a technology is plotted against competitive position, a graphical representation of potential threats and opportunities results. Referring to our coatings example, as shown in Figure 11-9, the need to upgrade competitive position in cross-linkers and metered spray equipment is reinforced. A new piece of information is also developed, which indicates that AAP's current colorimetry is more than adequate for the development of two-component enamels. The concept of shifting resources from programs with diminishing returns to programs with greater potential impact is introduced at this point. In this case, consideration should be given to shifting resources from colorimetry to one or both of the lagging technologies.

New monomers and pigmentation, as technologies, are relatively mature. This suggests that there is a low probability of identifying a new monomer or a new pigment that would provide speed of dry and acid rain resistance. By contrast, low molecular-weight components and cross-linkers are in their growth phase, and the chances are better of finding new materials in these technologies to address the desired attributes. Even though AAP is a leader in low-molecular-weight components the field is growing, and it would be worthwhile to include this technology in the search protocol. At this point we have three technologies worth searching: low-molecular-weight components and cross-linkers, as they relate to speed of dry and acid rain resistance, and metered spray equipment, as it relates to productivity and application latitude.

Competency versus Competitive Position.   Another consideration is the current level of competency that the organization possesses in each of the selected technologies, and the desired level to be attained in the future. Competency is not the same as competitive position. It is possible to have a low level of competency in a technology, yet be in leading competitive position. This paradox can occur particularly in an emerging technology. Conversely, in a mature technology, it is possible to have a high level of competency, yet still trail competition.

Figure 11-10 shows a matrix in which competency in each technology is plotted against competitive position. Two separate plots are first made. The first plot represents the current state of competency and competitive position. Upon analyzing one's current position and taking into consideration the most likely future scenarios, a second matrix can be developed representing the desired future state of competency and competitive position. When these two matrices are combined for each technology, vectors

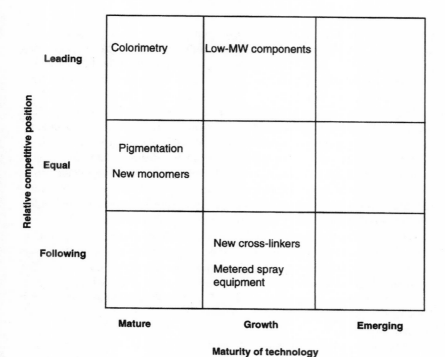

	Mature	Growth	Emerging
Leading	Potential waste of resources	Potential source of near term competitive advantage	Candidate for sustained investment
Equal		Industry average	
Following	Survival issue	Near term vulnerability	Potential future competitive threat

Relative competitive position

Maturity of component technology

Figure 11-8. Component technology opportunity chart.

	Mature	Growth	Emerging
Leading	Colorimetry	Low-MW components	
Equal	Pigmentation New monomers		
Following		New cross-linkers Metered spray equipment	

Relative competitive position

Maturity of technology

Figure 11-9. Technology opportunity chart. Two-component automotive repair enamels.

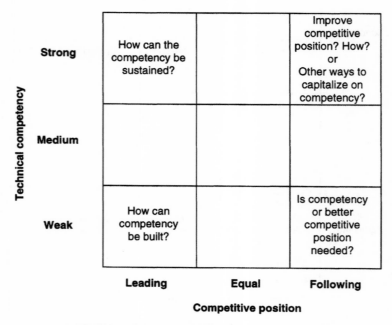

Figure 11-10. Technical competency versus competitive position.

can be developed, providing a graphic picture of the direction and size of effort likely to be necessary to achieve the desired state. These vectors can provide additional insight into the desirability and likelihood of achieving the desired state with internal or external resources within the desired time frame.

Again, let's return to our coatings example. Figure 11-11 shows the current level of competency and competitive position for the two-component enamel technologies. Current competencies correlate well with competitive position in this particular case. The pictorial view of the desired state is shown in Figure 11-12. Shifts in position are made only for the two lagging technologies, cross-linkers and metered spray equipment. In both cases AAP is realistic in their expectations of becoming competitive with these technologies, but not needing to strive for a leading position.

With regard to the level of competency desired, AAP is not an equipment manufacturer and does not expect to attain the same level of competency as the equipment manufacturers may have in metered spray equipment. On the other hand, cross-linkers represent a chemical technology which is potentially within the core competency of AAP, so that the goal of attaining a high level of competency in cross-linker chemistry somewhere in the future is a realistic possibility.

Using the charts for current and desired states, the direction and magnitude of these efforts are shown by the vectors in Figure 11-13. Metered spray

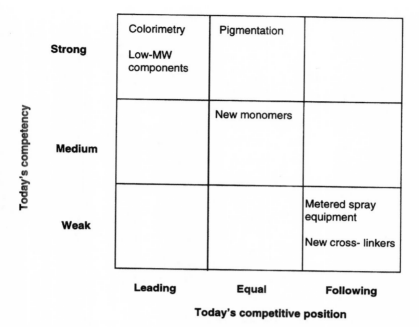

Figure 11-11. Current technology competency versus current competitive position. Two-component automotive repair topcoat enamels.

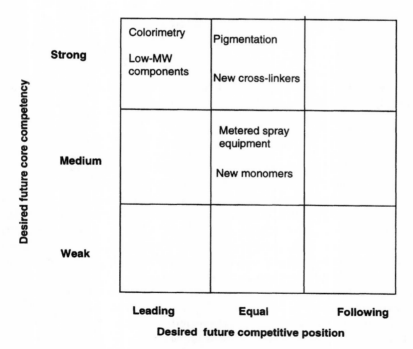

Figure 11-12. Desired future competency versus desired future competitive position. Two-component automotive repair topcoat enamels.

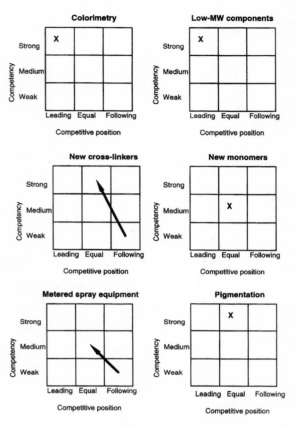

Figure 11-13. Vectors for competency and competitive position. Two-component autom7tive repair topcoat enamels.

equipment technology's vector represents a two-box move (one box up and one box across), while cross-linker technology represents a three-box move (two boxes up and one box across). This figure provides a visual idea of the magnitude of the desired change, and consequently some idea of the amount and type of resources that might be necessary to achieve the desired goals. Clearly, developing the desired position in cross-linkers will require a substantial effort in time and manpower compared to the effort required for metered spray application. The time element can be reduced through external technology acquisition, as opposed to internal development, but the demands for adequate manpower will probably remain the same regardless of the source of the technology. Before addressing the "how to", let's look at several more parameters, business impact and risk, as shown in Figure 11-14.

Impact and Risk.   The business impact/risk matrix can be used to classify technologies according to business impact and degree of risk, both technical and commercial. For these purposes, high business impact is defined as increased revenues and/or increased earnings. Low business impact is defined as maintaining the current position with regard to revenues and earnings. Risk is defined as the combination of probability of success and the costs that are involved. Therefore a high-risk technology could have a low probability of success and high costs associated with research and development. A high commercial risk may involve a low probability of success and significant barriers to customer acceptance of the new technology.

In the technology-selection/resource-allocation process, the ideal project would have high business impact and low technical and commercial risk. While such projects exist, often high-impact projects carry some sort of high risk, either technical or commercial, if not both. The concept here is to find ways to reduce these risks, so that the benefit can be derived at a reasonable cost to the organization in terms of time and resources, both financial and human.

There are a number of options for risk reduction. One option is to employ a phase-review system in which each stage of development is limited until key success criteria are met. Other options are to acquire the technology from external sources if it exists, or to share the risk with a partner through some

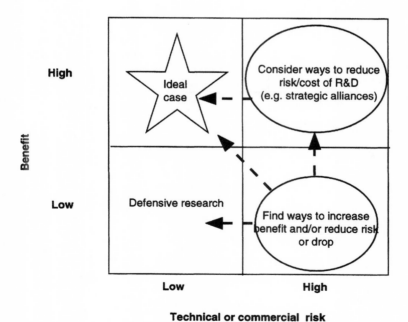

Figure 11-14. Benefit to business versus technical or commercial risk.

sort of strategic alliance, such as a joint development program. Risk management is an important activity within any organization, because if risk is simply avoided, it is likely that the organization will be limiting itself to low-
impact programs and a no-growth situation.

Returning to the coatings example, Figure 11-15 is typical for a relatively
mature industry, in which the technologies with low technical risk have low
impact, and high-impact technologies are associated with high technical risk,
in this case, the three technologies that have the best chance of improving
both speed of dry and acid rain resistance. The technical risk is high because
any change in the organic binder of the coating can affect all physical properties, and therefore requires extensive testing before commercial introduc-
tion can be initiated.

Metered spray technology presents an interesting issue. This type of
equipment is used to spray-apply two or more component systems which
react with each other once they are mixed together, and are therefore unstable

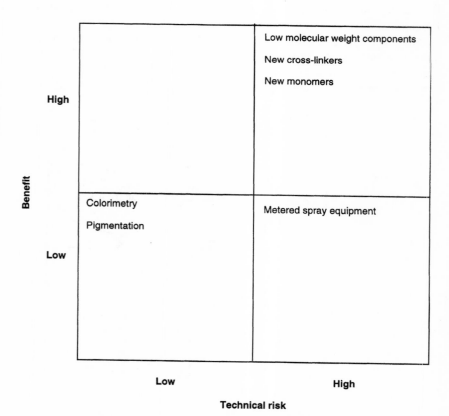

Figure 11-15. Benefit to business versus technical risk. Two-component automotive repair topcoat enamels.

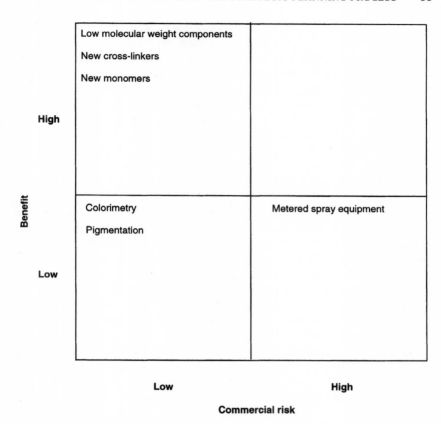

Figure 11-16. Benefit to business versus commercial risk. Two-component automotive repair topcoat enamels.

as a one-package system. The concern regarding technical risk is the question of equipment reliability and reproducibility in accurate metering of the multiple components over long time periods of use. The business impact of this technology is considered to be low, because AAP as a paint supplier does not foresee any differentiating competitive advantage. Metered spray equipment will be used only if absolutely required by the customer, therefore AAP has chosen a defensive strategy.

Figure 11-16 provides an analysis of commercial risk versus business impact. With the exception of metered spray technology, this chart confirms that the issues of risk are technical and not a matter of commercial acceptance. Commercial risk is high with metered spray technology, because if the customer does not already have this equipment in place, new capital investment will be required, as well as time spent on learning to use the equipment.

The issue of technical risk is one more reason to look to external technology for new cross-linkers and new monomers, and to consider low-molecular-weight components. An external partner, having more competency in these technologies, can help increase the probability of technical success while reducing time (and cost) to commercialization. For these reasons, many paint companies depend upon suppliers for these technologies.

Putting It All Together

In our illustration, the customer attribute/technology matrix showed that two-component enamels have the best chance of meeting customers' needs for the desired balance of properties, provided that the speed of dry of these systems can be improved. The technology-to-product concept matrix showed that cross-linkers, new monomers, and low-molecular-weight components have the best chance of providing the desired improvements in two-component enamels.

Although all of the technologies are important to the business in both the tactical and strategic time frames, the desired improvement in the balance of properties between acid rain resistance and speed of dry is needed within the shorter time frame of 2 to 5 years, as well as the need to become competitive in metered spray equipment. Given AAP's weak competitive position, with regard to both cross-linkers and metered spray equipment, external technology solutions should be sought.

The technology opportunity matrix for two-component enamels reinforces the need to improve competitive position in new cross-linkers and metered spray equipment. It also suggests shifting resources out of colorimetry and into these two technologies. In addition, cross-linkers and low-molecular-weight components may be the most productive areas to investigate because these technologies are still growing, while monomers tend to be more mature in terms of development.

The desired versus current technology competency matrices, show AAP's desire to substantially upgrade its competency in cross-linker and metered spray equipment technology, as well as its competitive position. A greater effort will be required in the cross-linker technology, while efforts in metered spray equipment technology will be more defensive in nature. In these instances, there is a direct correlation between competitive position and competency, although this is not necessarily always the case.

The business impact versus risk matrices confirm the importance of cross-linkers, along with new monomers and low-molecular-weight components. Given the technical risk involved in the successful development of cross-linker technology, the matrices suggest reducing risk through external partnering. This is also true with metered spray equipment, which represents high commercial as well as high technical risk.

Taking Action

Clearly, the strategic technology plan deals with all the technical resources available to the organization. The plan is comprehensive, covering issues that may be addressed by internal or external resources. Although this book focuses on external resources, they cannot be addressed in a vacuum, but only by integration with internal activities as well.

Taking action starts with a belief in the results of the strategic technology plan and the will to use it as guide to redeploy technical resources. Decisions need to be made regarding where internal resources will be focused, which technologies and programs are best out sourced and which programs should be downsized or eliminated.

Dealing with external technology initiatives begins with putting together a technology search protocol utilizing the five key sources outlined earlier in this book: patents, internal and external gatekeepers, lead users, and strategic suppliers. Subsequently, external technology initiatives, will involve redeployment of resources to these new programs, presumably from programs with diminishing returns, such as the Colorimetry program identified in the AAP example.

Technology Search.   Level 3 patent searches can be designed around each of the four technologies of interest; cross-linkers, monomers, low-molecular-weight components, and metered spray equipment. Through the use of appropriate key words, the goal is to determine who else in the world may be working in these technologies beyond AAP's direct competitors.

Internal gatekeepers can help design the patent searches through identification of key words. In addition, they should provide insights into which suppliers or organizations may be working in these technologies, and who might possess the highest levels of competency.

Activities such as key expert panels can be used to validate known external sources and possibly to identify additional sources to build upon the efforts of the internal gatekeepers.

Large refinishers who may already be evaluating new materials, may not only be a further source of information, but could represent potential customer evaluation and test facilities for experimental products and equipment.

Acquiring the Technology.   If the appropriate technology is available from current suppliers, a straightforward customer/supplier relationship will solve the problem. Where this is not the case, the technology will have to be either acquired or developed through some form of strategic alliance with an appropriate partner identified as a result of the technology search.

GUIDING PRINCIPLE

Principle #24   Linking competitive technical intelligence to a strategic technology plan is an ideal way to ensure an actionable and proactive process.

12

Acquisition of External Technology

Strategic Alliances as the Mode of Technology Acquisition

Having identified a noncompetitive source or sources of a desired technology, one might think that technology acquisition becomes a straightforward matter of simply purchasing the technology. Not today. It turns out that cash is the cheapest commodity when it comes to technology transfer. Companies that have invested resources to develop new technology are generally not interested in selling their investment for cash alone. More often than not, they either want a "piece of the action", beyond a simple licensing royalty, or prefer to receive technology of interest to them, in return. Technology has become too expensive to develop and too valuable to simply sell for cash or a royalty stream, unless it is considered truly nonstrategic in nature to the seller. As a result, acquisition of a technology usually occurs through some kind of strategic alliance, business merger, or acquisition.

Definition of a Strategic Alliance

The term "strategic alliance" has been used so much in recent years, that it has become a buzz word with fuzzy meaning. For our purposes, a strategic alliance is defined as any relationship between two parties to achieve a business purpose, that falls within the strategic window of opportunity time frame. As pointed out earlier, the strategic time frame for development of a new product, process, or service, is the time required to go from a clean-sheet design to commercial launch. In the chemical and allied industries, this time frame is generally five to ten years, or more.

Relationships that are developed for shorter time spans and for tactical purposes may represent an alliances, but not strategic ones. At the other end of the spectrum, mergers and acquisitions are not strategic alliances, because once formed, they no longer represent two parties, as they become a single unit.

Relationship to Business Strategy

Table 12-1 presents a hierarchy of strategic alliances listed in the order of increasing commitment between the parties, with a paid-up license representing the lowest level of commitment and a joint venture representing the highest. The appropriate level within this hierarchy should be aligned with the business strategy. Before approaching a party that possesses the technology being sought, an organization should have a clear view of how the acquisition of the desired technology fits into its own business and technology strategies. An assessment should be made ahead of time, utilizing the hierarchy table, of what level of commitment would be appropriate. It is wise to anticipate the demands of the technology holder in terms of their desired level of commitment, so that a fall-back position can be developed before contact is initiated.

Within the hierarchy shown in Table 12-1, as the level of commitment increases, so does the complexity of the relationship, as well as the level of risk. A key principle is to seek no higher level of commitment than is necessary to achieve the desired business purpose. The goal should be to estab-

Table 12-1 Hierarchy of Strategic Alliances (In order of increasing commitment)

1. Customer/supplier relationship

2. Paid-up license
3. Royalty-bearing license
4. Royalty-bearing license, with updates and grant-backs
5. Joint development
   - Supplier
   - Competitor in noncompeting segment
   - Customer
   - New entrant
   - Substitute product   Strategic
6. Functional joint venture   alliances
   - R&D
   - Manufacturing
   - Sales/marketing
7. Joint venture

8. Acquisition of the business

lish relationships that accelerate achieving business objectives at minimum cost and time. In those cases where a high level of commitment is desired, it should still be approached in a gradual step-wise fashion, if success is to be ensured. Without question, any strategic alliance is a process, not just a deal!

In one example, a working relationship was developed between a North American and a Japanese company. The relationship started with a simple licensing arrangement, which then grew into more complex licenses as mutual interests developed, and as the parties came to know and trust each other better. Subsequently, the two companies initiated joint development programs, and as business developed, cooperative marketing and sales arrangements were initiated. Finally, 17 years after the working relationship was first initiated, the companies formed a joint venture! The relationship has been a continuing process, with greater commitments built on a mutually successful business relationship.

In this relationship, the teacher was the Japanese company, with the North American company as the pupil. North American companies tend to be deal-oriented rather than process-oriented. North American companies tend to believe that the process is over once the deal is consummated, when in fact, the process has only just begun. The goal should never be to close the deal per se. In retrospect, the conservative, phased approach used in this example was also highly successful in terms of the competitive position of these partners relative to their North American and Japanese counterparts. The relatively slow pace of the evolving relationship provided the time necessary for the parties to work out problems in their relationship. Although the managing director of the Japanese company had a vision of where the joint cooperation was heading, the initiation of each new phase was simply based upon success in the previous phase of the relationship. The ultimate formation of the joint venture was not predicted at the outset by either party. Shortly before the decision was made to establish the joint venture, the managing director of the Japanese company said, "When we started, 17 years ago, I never envisioned our relationship would come so far."

Various studies have indicated that most alliances (including joint ventures) will eventually break up. The criterion for measuring success should be how well the alliance achieves the strategic objectives of the business, not how long the alliance lasts.

Role of Competitive Intelligence

With competitive technical intelligence as the tool to identify external technology of interest, as well as the potential sources of the technology, there is still a role for this tool in the process of seeking a strategic alliance. Returning to Michael Porter's model, shown again in Figure 12-1[1], it is critical to not only understand the competitive environment in which the organization is operating, but the forces that might change the nature of competition in the

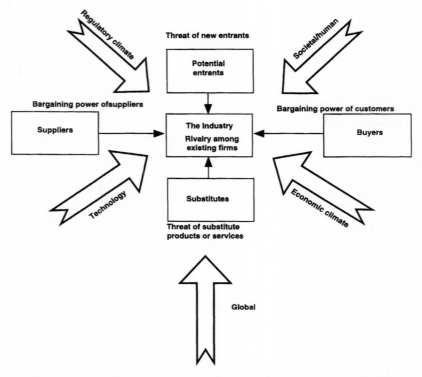

Figure 12-1. Key influences on technology trends, component technologies, and competitive position. Adapted from reference 1.

future. In today's environment, these issues need to be viewed from a global perspective, even if the organization operates in a limited geographic segment of the world.

If a strategic technology plan has been developed, these issues may already been defined. In either case, the four critical questions are: What does the strategic alignment of competition look like today? How might the strategic alignment of competition change in the future? Where does our enterprise stand today? How should we be participating in the future?

Determination of Strategic Intent

The most critical issue for the enterprise to come to grips with, prior to initiating discussions with anyone regarding a strategic alliance, is one's own strategic intent. Incompatible or poorly defined strategic intents are probably the major causes of failure of strategic alliances. There is one simple test on the adequacy of one's definition of strategic intent. Can the strategic intent be stated in one declarative sentence? If not, then it is poorly defined and requires more work before proceeding.

A Japanese supplier of automotive OEM paints to Japanese car manufacturers was seeking a U.S. partner prior to the Japanese car companies establishing manufacturing facilities in the United States. When asked what its intentions were, the Japanese paint company responded, "Our intention is to ensure a supply of high-quality paints to our good customers, the Japanese car manufacturers, anywhere in the world where they choose to build cars." Subsequently, this is precisely what the Japanese paint companies have done through global strategic alliances.

Boiling one's strategy down to a single declarative sentence is much more difficult than it might first appear. Organizations that have gone through strategic planning exercises have a distinct advantage in developing such an intent statement. Useful background information to develop this vision includes a good understanding of one's core competencies and sources of competitive advantage, as well as one's weaknesses and external threats to the enterprise. Using such information, one can then ask, "How should we be participating, given our business strategy?"

At the core of strategic intent is a good understanding of the definition of one's business. It is surprising how many companies struggle with their own self image, lacking clarity on who they are, who their customers are, and how they are perceived by these customers.

Once the issues of defining one's business have been resolved, and a strategic plan outlined, the task of developing a statement of strategic intent becomes much more straightforward. Why is there so much emphasis on strategic intent? If formation of the alliance will not clearly meet your strategic intent, do not proceed. Technology acquisition at any cost is not an acceptable option!

Mergers and Acquisitions

This book is not intended to cover the subject of mergers and acquisitions. It has been said that the planning for mergers and acquisitions focuses primarily on the financial considerations, and the process tends to take on the air of an auction. An advantage of mergers and acquisitions is that they tend to be consummated relatively rapidly compared to strategic alliances. There may also be an advantage of control by the acquiring company. On the downside, because they tend to be deal-rather than process-oriented (and rapid deals at that), mergers and acquisitions can result in incompatibilities that will ultimately result in their undoing. Less attention is likely to be given to strategic intents, technology, organizational compatibility, and corporate culture.

In situations where speed of action and control are the predominant issues, and organizational flexibility is sufficiently high to rapidly recover from "bad marriages," business mergers and acquisitions may be the appropriate choice to acquire new technology.

GUIDING PRINCIPLES

Principle #25: In a strategic alliance, seek no higher level of commitment than is necessary to achieve the desired business purpose.

Principle #26: A strategic alliance is a process, not just a deal.

Principle #27: The success of a strategic alliance should be measured by how well it achieves its business purpose, not how long it lasts.

13

Creating the Strategic Alliance: Partner Selection

Criteria for Partner Selection

For a specific technology of interest, there are usually no more than one or two potential partners available. Of course, this limitation simplifies the partner selection process.

Table 13-1 outlines the criteria for partner selection. With a view to the hierarchy of strategic alliances, the less commitment required by the parties, the less importance that can be placed on these criteria. The key exception is validating that the technology is, in fact, what is being sought. This is important regardless of the partners' level of commitment. By contrast, in most simple licensing transactions, the issues of the other party's strategic intent, organizational compatibility, and culture tend to be of lesser consequence.

These criteria start to become important in more complex licensing arrangements, when one is dependent upon one's licensor for updates, and is obligated to provide grant-backs. The criteria become even more critical if the intent is to engage in a joint development activity, let alone some form of joint venture.

Validation of the Technology

Much has been written on the subject of due diligence. One can obtain legal opinions regarding the strength of another party's patents, as well as a demonstration of the technology by the owner. The area of most confusion is

Table 13-1 Criteria for Partner Selection

1. Validated technology
2. Strategic intent
3. Organization compatibility
4. Culture

usually in regard to the level of development of the technology. One needs to be clear on what is being sought, and delivered. Is it technology in its early stages, or is it a fully developed product, process, or service, or something in between? People will talk about technology when, in fact, they are seeking a product, process, or service, and are subsequently disappointed to discover that what they got is not what they wanted.

The same applies to joint development efforts. It's important to understand the current state of development of the technology so that a realistic assessment can be made of the technical and commercial risks, as well as developing a realistic estimate of the time and resources necessary to carry the program through to commercial success. Given the state of the art, does one have the necessary competencies to bring the development to a successful conclusion? What value does each partner bring to the joint effort?

A small foreign company had developed a radical new process for curing coating systems, which, if proven effective, could substantially reduce manufacturing costs in painting operations. Having limited resources for global expansion and needing a cash infusion, the company was eager to grant licenses overseas. A U.S.-based company saw the potential for the technology, but did not have a clear picture of the current stage of development, nor of the technology's ultimate suitability for the industries it served.

The U.S.-based company resisted pressures from the would-be licensor to accept a complex license arrangement, and instead purchased a one-year option for future exclusive U.S. rights to the technology. This option gave the U.S. company a full year to evaluate the new technology. As it turned out, the technology had a number of previously unrecognized limitations and was not suited for the industries addressed by the U.S.-based company, and so a license was never consummated. The option fee turned out to be a wise expenditure, providing time to validate the technology, while limiting the potential licensee's commitment.

Strategic Intent

The subject of strategic intent was covered in Chapter 12. The test can now be applied to determine if a prospective partner has thought through its strategic intent. Can the partner state its strategic intent in one simple declarative sentence? If not, caution is indicated. The degree of commitment should be limited, until such time as the prospective partner can clearly state its intentions.

In the example of the Japanese and North American companies that took 17 years to commit to a joint venture, the rate of development of this relationship was controlled primarily by the Japanese company. For many years, the North American company was struggling with defining its strategic intent. Once both companies were able to clearly articulate their intent, the companies moved rapidly to establish the joint venture. The Japanese company admitted, subsequently, that it was delaying considering a joint venture, because of its partner's inability to clearly articulate its strategic intent.

Once a potential partner can articulate its strategic intent, one must evaluate the compatibility of the goals of the two enterprises. Table 13-2 shows the three possibilities that can exist regarding compatibility of goals.

The goals of two parties, in alignment, are not necessarily the same. They can simply be compatible. In the case of the Japanese and North American companies, the strategic intent of the Japanese company was to ensure a supply of high-quality paints to its good customers, the Japanese car manufacturers, anywhere in the world it chose to build cars. The strategic intent of the North American company was to be the leading supplier of high-quality paints to the Japanese car manufacturers in North America, in cooperation with the Japanese paint company. These goals are different, but compatible and aligned with each other.

When goals are in opposition, this may not necessarily be a deal breaker, but the caution light is on. For example, car manufacturers and other powerful customers may require a supplier to license their technology to a competitor to assure the customer of two sources of supply. In such cases, alliances should be limited to arms-length transactions, such as licenses, but care needs to be taken to assure technology transfer and adequate provision for updates.

In one such case, a supplier agreed to license a competitor, but the subsequent technology transfer was inadequate and no provision was made for updates. The licensee was placed in a position of severe competitive disadvantage, having continued difficulty in managing the licensed technology as a supplier to its mutual customer. Ultimately, because of poor technical service and substandard product performance, the business awarded to the licensee was withdrawn and returned to the licensor.

In the case of unrelated goals, the two parties have different objectives that are not linked to each other, and are seeking different benefits from an

Table 13-2 Compatibility of Strategic Intents

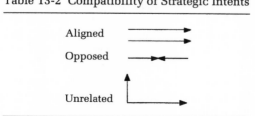

alliance. This may or may not result in a compatible relationship, depending upon the specific goals. It is critical to understand the intents of the other party and to have a shared view of how each party's needs and wants will be met through the alliance.

An example of unrelated goals might be the GM-Toyota joint venture in manufacturing, New United Motors Manufacturing Inc. (NUMMI), in Fremont, California. GM's intent was to learn Toyota's manufacturing and quality-assurance practices, while Toyota was intent on learning how to operate in the U.S. labor market. Both parties benefited while successfully producing quality cars under the Geo Prizm and Toyota Corolla nameplates, and marketing their vehicles separately. Although unrelated, their goals were compatible.

Another measure of the compatibility of strategic intents is to assess the value that will be created as a result of the alliance. Is the value greater than what can be achieved by the partners operating alone? What value does each partner bring and take away?

Organizational Compatibility

The issue of organizational compatibility becomes increasingly important with increasing level of commitment within the strategic alliance hierarchy. Even in a reciprocal licensing relationship, such as in an update and grant-back framework, organizational compatibility comes into play to some degree.

One needs to understand the other's organizational structure, and how it operates. Where are the key decisions made in the structure? Who are the appropriate people to be interfacing with in the partner's organization?

European and Asian companies tend to keep key people in the same positions for longer periods of time than in U.S. companies. Many of these foreign companies have expressed frustration in dealing with U.S. companies because of turnover of their key contacts and decision makers during the life of their strategic alliance. This turnover results in the remaining partners having to bring the new people up to speed and also carries the risk of a lack of consistency in organizational behavior, and possible differences in interpretation of the original strategic intents between the parties.

As silly as it may sound, in one case, when I was seeking a license from a European company, the licensing director furiously exclaimed, "This is the third time your company has requested a license, and each time it has come from a different person. I never know who I am dealing with." The request for a license was refused. Several months later, I called again, pointing out I was the same person who had called previously. We subsequently got the license.

In another alliance with a Japanese company, the Japanese R&D director had remained in the same position for the 12 years that the alliance had existed. In this time period, the U.S. company had 5 different R&D directors move through the organization. As each new R&D director was introduced to his Japanese counterpart, the Japanese director, in frustration, would recite the names of all the U.S. director's predecessors, saying that he hoped that the new director would stay a while in his position.

Recognizing this turnover to be a serious issue, another U.S. company became committed to maintaining continuity at the interface with a partner engaged in joint development programs. Over a period of 15 years, the U.S. company kept the same individual as its key interface with its partner, even though this individual's job changed several times during this time period. The continuity has been appreciated by the partner, who has also worked at minimizing turnover at the interface. Both parties consider their joint program a model for success.

Culture

With increasing commitment and complexity in the strategic alliance, cultural issues also become increasingly important. In dealing with global strategic alliances, there are many differences between peoples from the various continents, as well as differences within these groups.

North Americans tend be deal-oriented, while the rest of the world is more sensitive to the need for process in negotiations. Some years ago, we were expecting a visit from a vice-president of a major Japanese company. In arranging for the visit, the vice-president requested about two hours of our time, but offered no formal agenda. Our senior management was distressed. What was the purpose of the Japanese vice-president's visit? What could we possibly productively discuss for two hours?

No sooner were we seated in the conference room, when our vice-president asked of our visitor, "Sir, what is the purpose of your visit?". Our visitor replied, "In Japan, there are three phases to a business relationship. First, it's a matter of getting to know you. Secondly, we try to identify areas of common interest. Third, we seek agreement to cooperate in one or more of these areas. Today, I am here for the first phase, I've simply come here to get to know you." With this statement, he completely disarmed his hosts. Everyone relaxed and proceeded to have a productive discussion, sharing views of the industry from a global perspective and the dynamics of the marketplaces served by the two companies. Over the ensuing years, the two parties eventually consummated several licensing agreements.

On another occasion, one of our vice-presidents requested that we seek a technology license from a European company with whom we had been carrying on a running dialog. When told that a visit to discuss the license would be arranged, the vice-president exclaimed "No need! Why not just give them a call?" When the phone call was put through to the European contact, and the request for the license was made, the European colleague replied, "Yes, but where's the romance?" We replied, "We'll be over next week." We subsequently got the license we sought.

As U.S. companies become more global in their perspective, there is growing recognition that strategic alliances are a process, not a deal, so that U.S. companies are closing this cultural gap with their partners overseas.

There is a substantial difference in how agreements are viewed by Japanese and Westerners. Westerners see the *agreement* as the key to the alliance, while Japanese see *relationships* as the key to the alliance, with the agreement simply a historical record of what the two parties have previously agreed to. The Japanese approach allows for more flexibility within the relationship and the alliance, as well—an important attribute necessary to deal with a changing environment and changes within the industry. Again, there is growing awareness within the western community regarding the importance of relationships and flexibility in any strategic alliance.

Finally, recognizing that many cultural differences exist between people all over the world, the key question is, "How can we bridge cultural differences to facilitate working together?" First comes awareness that cultural differences exist, and then comes the willingness to understand these differences. Books and training programs, dealing with most cultures, abound. When training and education run out, politeness and professionalism will invariably carry one through.

On a visit to Japan, we were introducing our new R&D director to his Japanese counterpart. We had a long relationship with this particular company, and up to this point had worked very closely with them on a number of projects. After dinner, the Japanese R&D Director had each participant, Japanese and American, sing his college *alma mater*. Our new R&D director

---

GUIDING PRINCIPLES

Principle #28:   Due diligence is critical to understanding the stage of development of the technology, product, process, or service to be acquired.

Principle #29:   Compatibility of the parties' strategic intents is key to a successful strategic alliance.

Principle #30:   Relations are important; maintaining continuity in interface personnel goes a long way in providing stability to a strategic alliance.

Principle #31:   Strategic intent must not be compromised for the sake of the relationship if the alliance is to succeed.

sang with great enthusiasm, albeit off-key, when his turn came. The Japanese director turned to me and said, "I can work with this man."

Having recognized the importance of organizational compatibility and culture, it is important to keep in mind that in selecting a partner, one must not compromise one's strategic intent for the sake of the relationship.

14

Negotiating the Strategic Alliance

Planning and Organizing for the Negotiation

Generally, U.S. businesses do an inadequate job of planning their negotiations, when compared to their foreign counterparts. As a licensing executive, one of the more frustrating experiences is to get one's own key executives to recognize the steps involved prior to negotiating with the other party, as outlined in Table 14-1. There is a propensity to immediately initiate contact with the source of the desired technology, before any planning has occurred. What needs to be recognized, is that when that first contact is made, *negotiations have already begun!*

A schematic organizational arrangement to deal with alliance negotiations is shown in Figure 14-1.

Selecting the Negotiating Team. The negotiating team is a group that evolves with time. Membership can vary depending on the stage of negotiation and the level of commitment involved in the strategic alliance. The initial phase is usually a scouting mission to determine the interest of the other party in starting a dialog. The scouting team can consist of as little as one person, if that individual has an adequate knowledge of the business interest, technology and technology transfer, and cultural and language background, if appropriate. Alternatively, additional members may be added to the team to fulfill one or more of these roles.

It's most effective for the negotiating team to report to a technology transfer board (TTB), which consists of senior executives who have power of approval. Usually, the TTB is a cross-functional group including directors or

Table 14-1 Planning the Negotiation

---

1. Select the negotiating team
2. Select the support team
3. Reiterate the strategic intent
4. Develop negotiation strategy
5. Identify roles and ground rules for negotiating team members
6. Clarify negotiating tactics
7. Initiate contact

---

vice-presidents for the business, R&D, manufacturing, finance, and legal functions. The TTB would be responsible for approving the negotiating plan, thereby empowering the negotiating team to act, while at the same time establishing the boundary conditions, beyond which the team could not go without further approval of the TTB.

Selecting the Support Team.   The support team is a cross-functional group with appropriate skills and knowledge, brought together to advise the negotiating team. The support team can be a small, informal group for simple

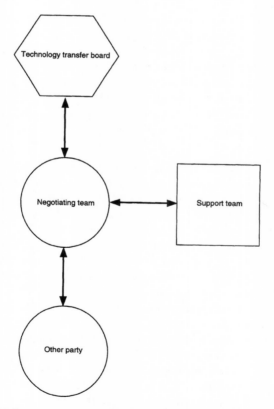

Figure 14-1. Schematic organization for the negotiating process.

licensing transactions, or a large, formal team for more complex alliance negotiations. Such a group can play several roles, such as providing expert advice for negotiation planning purposes, or it can be a sounding board for the negotiating team. Individual support team members may be called on for more in-depth briefings, and may be brought into the negotiations as deliberations progress.

Strategic Intent and Negotiation Strategy.   Strategic intent and an outline of negotiating strategy should be developed in writing. An example of a practical worksheet is shown in Figure 14-2.

Once the negotiating team has reaffirmed the statement of strategic intent, it needs to itemize its "wants". In simple terms, just what does the organization want from the alliance? What does the other party want? What costs does the organization intend to put into the alliance? What will the other party put in?

Issues of conflict between the parties can now be identified. What does one's organization see as the issues? What will the other party see? One should not assume that the issues will be the same for both parties. The issues can be different.

Before proceeding to deal with the question of a fall-back position, it is important to revisit the list of outputs and inputs, and to prioritize them into "musts" and "wants" for both parties. Once this has been done, one can proceed to develop a fall back-position based on these musts and wants.

This basic document can also serve as a communication device with the technology transfer board to facilitate its authorization. Prior to seeking authorization with the TTB, a review with the support team will help to clarify issues and gain support for the strategy as well.

Roles and Ground Rules.   With a written strategy in hand, the negotiating team can then deal with roles for the various team members and ground rules regarding how the team will operate. Who will be the primary speaker? When is it appropriate for others on the team to speak? Who will take notes? Who will be the process observer and resource for the team? Will the resource be empowered to call for a "time-out" for a private caucus? How will time be managed? Is the team prepared to "walk away" if their "musts" cannot be satisfied? What does the team do if an impasse is reached? What is the best alternative to a negotiated agreement?

The importance of preparation for negotiation cannot be overstated. Foreign organizations tend to do a much better job of preparation than those in the United States, because the foreign organizations more readily recognize that negotiations are a process. The negotiation worksheet, just described, enables the team to develop alignment around its objectives, and anticipate issues and points of difference. It is often the case that the internal negotiations are more difficult than those with the other party.

Strategic Alliance Negotiation Worksheet

Strategic intent:_____

_____

_____

We_____**Output-wants**_____**They**_____

We_____**Inputs-costs**_____**They**_____

We_____**Major issues**_____**They**_____

We_____**Fall-back position - musts vs. wants**____**They**_____

Figure 14-2. Strategic alliance negotiation worksheet.

Negotiating Principles

Although some people seem to have natural negotiation skills, these skills can be developed. Courses are available which provide the participant with opportunities to not only plan negotiations, but to engage in role-play as well. The first principle of negotiation is that the "deal" has to be a win-win event. Culturally, Americans tend to think win-lose when negotiating, and tend to

Table 14-2 Principles of Negotiation

1. Plan the initial contact—negotiations have begun with first contact!
2. Start with high aspirations
• Aspirations can always be reduced, but hardly ever raised
3. Keep in mind each party's power position
4. Stay in control—take the initiative
• Prepare the documents
5. Concede slowly
6. Don't be reluctant to take a break for team caucuses in private
7. Deal with impasse through innovation
• Be prepared to "enlarge the pot"
• Always know best alternative
8. Don't get locked in to doing the deal
9. Don't compromise strategic intents for relationships
10. Make time an ally, not an enemy
11. Perseverance is key
• Patience! Patience! Patience!
12. Continuously assess state of negotiations against value
13. Walk away if deal becomes unattractive

think of win-win as a sign of weakness. Both parties need to come away from an agreement with a good feeling about the agreement. If a party nurtures bad feelings, they may not get mad, but they will find ways to get even. In no way does this imply the need to go to the extreme of trading substance for relationships or good feelings. Table 14-2 outlines 13 key principles in conducting negotiations.

The Initial Contact.   It can't be emphasized enough that negotiations begin the moment first contact is made. One can rationalize that the initial contact is simply a scouting mission, having no impact on the final outcome, but this is only rationalization.

A company (party A) engaged in a major new technology development discovered that another company in another part of the world (party B) was similarly engaged. Perceiving this activity as a competitive threat, party A sought to "tie-up" party B's technology by seeking an exclusive paid-up license. At the outset, party A failed to consider what the incentive would be for party B. Upon contacting party B, and making the request to purchase the technology, party A was surprised at the hostile rejection. "Why should we sell out 12 years of research?" queried party B.

Once party A realized its strategic error, it tried to make amends by offering the other party participation in an alliance. It was too late. Party B, recognizing the commitment party A had to this technology, believed that party A's intent was to simply tie up the competing technology and patents to improve its own competitive position. As a result, party B broke off contact with party A, and subsequently entered into a strategic alliance with party A's competitor.

Who will make the first contact? Will it be the CEO or vice-president, or a middle-or first-line manager, or a technical professional? Will contact be made by letter, e-mail, phone, or fax? Will the purpose of the contact be revealed immediately, or will it be held back until a personal visit is arranged? Who will be involved in the introductory visit?

Start with High Aspirations.   "You've got to have a dream. If you don't have a dream, how are you going to have a dream come true?" goes the song from the show, *South Pacific*. And so it goes in negotiations. Aim high. One can always come down, but one can almost never go up. Common sense would indicate this to be true in virtually all human transactions involving "bid" and "ask" terms.

Power Position.   In negotiations it is easy to fall into the trap of developing unrealistic views of the power accorded each party. Human nature tends to accord too much power to the other party, which can unduly weaken one's perception of its own bargaining position, and can end up with suboptimal results.

In other cases, an organization can overstate the power of its own position, resulting in unrealistic demands in the negotiation and a consequent loss of credibility and bargaining position. On one occasion, a company put such a high value on a trademark that it was no longer using, that the other party withdrew its request to purchase the trademark. Years later, the same trademark was sold to the same party involved in the earlier discussion, at a substantially lower price.

Because of the potential trap of becoming too subjective regarding power positions, it is wise to utilize an external sounding board to make this assessment. Here's one area wherein a support team can prove useful.

Stay in Control.   Maintaining control of the negotiations increases one's chances of successfully attaining one's goals. The best way to maintain control is to take a proactive approach to the negotiations. For example, volunteer to prepare any agreements and documents required. The initial documents, such as a letter of intent or heads of an agreement, can be most important in terms of the direction the deliberations will take. Let the other party react to the drafts of these agreements, rather than the other way around.

Similarly, take the initiative in setting the location. In a negotiation, a company chose to set the venue at its headquarters location. The other party arrived from overseas with a large delegation. During the subsequent negotiations, an impasse was reached, and the host organization walked out, leaving the large contingent of visitors stuck far from home. Subsequently, the visitors gave in to the key requirements of the "home team" rather than return home empty-handed. The negotiations may have had a significantly different outcome had the venue been reversed, or had the size of the visiting contingent been substantially smaller, whereby it would have been easier for the other party to walk away.

Concede Slowly.   Being proactive and maintaining control does not mean that all actions need to be quick and decisive. Given the principle that one can always come down in terms of demands, but rarely can one increase them, it follows that concessions should be made slowly. This gives the team the opportunity to test the resolve of the other party.

A company seeking a license from another party, received a counter request from the other party for a down payment twice the size offered by the prospective licensee. After a caucus, the prospective licensee quickly counteroffered to make a down payment halfway between the two figures. This counteroffer was agreed to by the licensor. Later, over cocktails, the license learned that the licensor was willing to come back down to the original down payment offer if the licensee had not so quickly offered a compromise!

Caucus.   Many Americans seem to think it is a loss of face, or a sign of weakness, for a negotiating team to break off negotiations and leave the room in order to caucus in private. Representatives of foreign companies will often caucus in the room where negotiations are taking place, in front of the other party, in their native language, if they are comfortable that the other party doesn't understand that language. Because Americans don't have that advantage, it is necessary for them to leave the room to caucus.

Caucuses are in order when new information or a new request is introduced by the other party, that wasn't anticipated by the negotiating team. If the role of a process resource person has been established within the team, it is usually this person's responsibility to recognize when a private caucus is in order. Nevertheless, anyone on the team can call for a private caucus. A key ground rule is that when a caucus is called for, it should receive top priority and be acted upon immediately.

During a negotiation, the spokesman for one company became angry over a request made by the other party. Because the leader's anger was apparent, and his reasoning powers impaired, a team member proposed a recess to caucus. The leader rejected the proposal and continued to express his anger, resulting in compromising the team's bargaining position.

Never assume that the other party does not understand English. Embarrassing moments can occur when a negotiating team decides to caucus in front of the other party. Particularly in dealing with companies from the Far East, there is likely to be one translator who speaks English while the rest of their team remain silent or only speak in their native tongue. On one occasion, a team I was working with chose to caucus in front of their Japanese counterparts. When we had concluded our open discussion, one of the Japanese men, who had not spoken a word all morning, said, in perfect clipped English, "How very convenient for you." We had, unwittingly, just exposed our "hand" to the other party.

Dealing with Impasse.   If impasse appears to be reached, and it looks like both parties are about ready to walk, it's time to get innovative regarding the scope of the alliance. The guiding principle here is that enlarging the scope of the alliance will usually overcome an impasse.

In a particular negotiation, an impasse was reached regarding the financial considerations for a license. The prospective licensee broke the impasse by offering to "throw in" some of their own technology into "the pot" for cross-licensing purposes. In another case, a U.S. company was seeking rights for an exclusive license from a Japanese company in North and South America and Europe. The Japanese firm did not want to relinquish rights in Europe. The impasse was broken by the parties agreeing to establishing a joint venture in Europe at some later date, thereby protecting both parties' European interests.

Doing the Deal.   Although not expressly stated, some people fall into the trap of believing that not doing the deal represents failure, and conversely, doing the deal represents success. If the desire for closure on an alliance overrides meeting the musts and wants of the organization, the results can be disastrous.

The negotiating teams from two global companies felt great pressure from their respective senior managements to get closure on negotiations to establish a joint venture company in another part of the world, despite the fact that the two companies had incompatible strategic intents. The teams gave in to this pressure, and completed the deal which established the joint venture. The resulting company was unsuccessful, and subsequently one partner sold out to the other.

Strategic Intents versus Relationships.   The guiding principle in a negotiation is to prevent the desire for harmony from compromising one's strategic aspirations. This is often easier said than done. There is a chemistry between the parties that can develop during a negotiation, whereby individuals say to themselves, "I can work with these people." Also, one's desire to be liked can creep into the picture.

At this point, the focus can shift from attaining one's goals to preserving relationships and not offending the other party, while rationalizing that differences regarding issues can be resolved once the alliance is in place. This is simply not so. As a beer commercial once said, "It doesn't get any better than this." Once the alliance is established, it is hard work to maintain the same level of relationship that was established during negotiations. Therefore, it is imperative to resolve issues during the negotiation phase, even at the risk of jeopardizing the relationship. If the relationship cannot survive dealing with the issues, it would not have been a good business relationship anyway.

Time as an Issue.   The guiding principle here is to understand that being dominated by time pressures can substantially weaken one's negotiation position, while mastery of the time element can become advantageous in negotiations. Americans tend to be deal-oriented to begin with. If this orientation is coupled with a desire to close on the deal, with some sort of deadline in mind, the results can be disastrous.

A small company engaged in licensing discussions was eager to receive a substantial lump-sum payment for a technology license in time to have the income reflected before the end of the current quarter. Once the other party became aware of the importance of the time element to this company, it was able to extract concession after concession from them, which the small company subsequently regretted. Ultimately, the small company felt that it had given away more value than the lump-sum payment was worth, just to meet its arbitrary deadline. Patience would have been a more profitable strategy.

Perseverance Can Be the Key to Success.   There are two aspects to perseverance. The first aspect, as described previously, is patience, which can be the key to success if one resists the pressure of time. If a strategic perspective is adopted for the negotiation, one can realize how unimportant the delay of weeks or months may be in terms of reaching a satisfactory result.

The second aspect of perseverance is simply not taking no for an answer. This aspect works as well in technology acquisition as it does in other aspects of life. It worked in the case, previously described, where the potential licensor complained that every time our company had approached him, it was a different person. By my approaching him a second time, I was able to overcome his objection. There have been other occasions where one may be "brushed off" with the comment that it is too premature to discuss licensing. Such a statement sets the stage for follow-up, time and again, until it becomes timely to discuss the subject.

For action-oriented Americans, accustomed to doing the deal and getting closure, the time lines involved in technology acquisition seem like an anathema. Today, mergers and acquisitions have become a popular alternative method to acquire technology, along with the entire business. One reason given in favor of this approach is that it operates much like an auction, and therefore is much faster than the courtship process of a traditional alliance.

Assess the State of Negotiations.   As negotiations progress, they can take on a life of their own, with the potential level of commitment increasing beyond the level originally anticipated. The principle here, is to periodically step back from the negotiations to assess the current state against the value to be extracted. Has the value been compromised too much by the costs associated? These costs could be financial, or involve more commitment than the organization is willing to make. For an objective assessment, it is worthwhile to periodically utilize the support team as a sounding board.

Walk Away.   The idea of walking away, if the deal turns unattractive, might be inconceivable if one has become locked in to doing the deal. The same principle applies for those who perceive that not doing the deal is failure.

The guiding principle here is to walk away if the deal has turned unattractive, with no chance of turning it around, but to leave the door open. Just as with the example of the overvalued trademark, as time went on, the would-be licensor became more realistic, and ultimately a license satisfactory to both parties was consummated.

Providing for Technology Transfer

Effective technology transfer is a process unto itself, but provision to enable successful execution of the transfer has to be made in the agreement(s) covering the alliance. Usually, the terms for the technology transfer are included in these agreements, so that the negotiating team needs to be conversant with these terms. This awareness may require adding different capabilities to the team as these negotiations become timely. Inadequate provisions in the technology transfer framework can destroy the value of the technology acquisition. The technology transfer needs to be complete and thorough.

GUIDING PRINCIPLES

Principle #32:   Negotiations have already begun when the first contact with the other party is made, regardless of how trivial the contact may seem.

Principle #33:   Preparation by the negotiation team is of paramount importance.

Principle #34:   Regarding strategic alliances, negotiations need to be structured for a win-win result.

15

Executing the Technology Transfer

Elements of Technology Transfer

The four major elements of technology transfer are outlined in Table 15-1. They are: the technology manual; in-person briefings, coaching and counseling to facilitate ramp-up and commercial launch, and subsequently, technology updates if the alliance embodies a license with running royalties. All four elements are critical to a successful technology transfer, and the time required to execute them is, more often then not, underestimated by the conveyor of the technology. As the recipient, it is therefore critical to understand what goes into these elements, so that negotiations provide for sufficient time allowance to adequately execute them.

Technology Manual. For the chemical industry, the contents of the technology manual would look like the outline shown in Table 15.1. The elapsed time required to pull all this information together, and translate it from company coding and jargon into plain English, is invariably underestimated. In my experience, the minimum elapsed time to document the simplest technology transfer is three man-months. More complex technologies will take longer.

Editing and testing the resulting manual are rarely done, although they should be included important steps. The final test is to give the manual to someone within the organization who understands the field of endeavor, but who is not intimately familiar with the specific technology. If that individual can clearly understand how to reproduce the results obtained by the technology provider, the manual is ready for conveyance to the recipient.

Table 15-1 Elements of Technology Transfer

---

1. Technology manual
   - Composition formulas
   - Manufacturing/process directions
   - Specifications
     Raw materials
     Intermediates
     Finished product
   - Test methods
   - Troubleshooting guide
   - List of relevant patents
2. Face to face briefings and consultation
   - Question and answer sessions
3. Ramp up coaching and counseling
4. Periodic updates

Briefings.   Invariably, regardless of the clarity and thoroughness of the technology manual, the receiving organization will have questions that require a face-to-face session or two to clarify or amplify on the written word. Allowance should be made for such sessions in any technology transfer agreements. Again, this step tends be overlooked or underestimated.

The recipient organization should be alert that provision is made for at least two briefing visits at the most convenient location for the recipient, and that there is clarity regarding who will bear the costs for the time and travel of the information provider(s). The rationale for at least two visits is that it is unrealistic to think of all the questions and contingencies during one visit. Allowance for a second visit is, therefore, an insurance policy. Naturally, the provider will strive to put some specific limits on these consultations, with appropriate fees and expense reimbursement for consultations that exceed agreed-upon limitations.

Ramp-up Consultation.   As process scale-up occurs, it is likely that glitches will arise that could not be anticipated by documentation and discussion. To help cope with unexpected problems, the recipient organization needs to make provision to have a consultant from the provider available. Again, at least two visits should be anticipated for even the simplest technology transfer, because it is unrealistic that every issue will be adequately covered in one visitation.

Technology Updates.   As a technology licensee, it is fair game to request and negotiate periodic updates. It may be more difficult to obtain these updates when the technology is purchased outright for a lump sum, but it doesn't hurt to try. A stronger argument can be made by the licensee who is paying a running royalty over a period of years for a technology license.

It is reasonable to expect the new licensee to take some significant time period to reach the same level of competency as the licensor, with regard to the acquired technology, and that the licensor will have momentum to continue to develop the new technology at a faster pace. Technology updates will enable the licensee to keep up until such time as they are comfortable to go it alone.

As described earlier, one of the more disastrous licensing arrangements was the result of the licensee not having access to technology updates, thereby preventing the licensee from staying competitive. If the licensor agrees to providing updates, they will usually expect to receive royalty-free grant-backs on further improvements on the technology made by the licensee. The licensee should anticipate this request when asking for update provisions, and needs to first assess the relative future rate of technology development and growth in competency of the two parties.

Organizing for the Technology Transfer

The steps involved in organizing for the technology transfer are outlined in Table 15-2. Often, the need to organize to receive technology is not recognized by the recipient, so that the transfer is often carried out in an ad hoc fashion. In time, it will be important to know who, within the organization, has possession of the newly acquired technology, that the technology transfer was complete and thorough, and that an organization exists in which to receive subsequent consultation, and periodic updates, if called for.

If the alliance involves some sort of ongoing joint development effort, then the need for a formal technology team is all the more imperative.

A formal technology transfer team needs to be defined. The members should represent all the skills and knowledge necessary to receive the technology and incorporate it into the organization's system. Two specific roles need to be defined: a team leader and a technology gatekeeper. The team leader will be required to have strong interpersonal skills, and is responsible for carrying out the administrative duties. The technology gatekeeper will be the primary source of information, as well as the technical continuity on the team. It is possible for both roles to be fulfilled by the same person.

Table 15-2 Organizing the Technology Transfer

---

1. Establish a technology transfer team.
2. Appoint a team leader and technology gatekeeper, with an eye to longer term continuity.
3. Minimize turnover of the team leader and technology gatekeeper.
4. Provide team members with written copies of the purpose of the alliance, goals, strategies and programs, as well as terms of the license agreement.
5. Establish a system for ongoing communications, periodic meetings and reviews.
6. Upon receipt of technology, make as few changes as possible as it is incorporated into the system.
7. Pursue necessary improvements and technology support programs aggressively.

The rationale behind establishing these roles is to minimize the turnover of the team leader and the technology gatekeeper. This is especially important where long-term continuity would be advantageous, such as in joint programs or other activities involving an ongoing dialog with the other party.

Team members should have written copies of the license agreement, as well as written statements regarding the purpose of the alliance, including goals, strategies, and programs. A system needs to be put in place for ongoing communications, periodic meetings, and reviews with the other party. There needs to be a single point of contact to take official action regarding such matters as program direction. Usually these are the team leaders. At the same time, it is important that there are lines of communication directly between the technical people, on a day-to-day basis.

In the best joint programs, technical people are free to communicate as needed directly with their counterparts. As a minimum, programs are reviewed by team leaders and their technology gatekeepers on a quarterly basis. Once a year, the two teams meet for a major review to assess progress, identify new opportunities, and to decide how their respective resources will be deployed for the coming year. Of course, a key consideration is whether to continue the alliance on the current basis.

Traditional wisdom dictates that upon receipt of the technology, the recipient organization should make as few changes as possible (preferably none) as it is adopted into the new organization. This is a good guideline to follow. However, in many cases the technology received may still be evolving or in need of modification to make it commercially viable. In such cases, it is important to aggressively pursue the necessary improvements, if competitive advantage is to be sustained.

On too many occasions, a technology is licensed in and then allowed to languish so that competition is given adequate time to catch up, and the value of the technology acquisition is substantially diminished. In other instances, the acquired technology is commercialized or may already be commercial when received, but no sustained improvement programs are in place. In time, the acquired technology becomes obsolete, and because the enterprise has not supported a "next generation" program, competitive position is eroded. Providing for a formal organization at the time of the technology transfer can safeguard against such neglect and erosion.

GUIDING PRINCIPLES

Principle #35:   Understanding the elements of technology transfer can help safeguard against the common problem of underestimating the time and effort required to effectively transfer technology.

Principle #36:   A key to effective technology transfer is the formation of a technology transfer team by the receiving party.

Principle #37:   Acquisition of external technology saves time only if it is aggressively utilized once acquired.

16

Structuring and Managing Strategic Technology Alliances

Structuring the Technology Alliance

A strategic technology alliance can begin with the structure used to facilitate the technology transfer. Ultimately, the amount and type of structure required will be determined by the level of commitment of the alliance described earlier. The minimum roles to be fulfilled are the team leader and technology gatekeeper previously discussed. These two roles provide the structure and pathways to manage and foster partner contacts and exchanges.

Once the strategic intents of the two parties are aligned, the next most critical issue is to staff the key roles with the right people. Relationships become the shock absorbers in an alliance. They absorb the blows from errors in behavior and judgment that characterize all human transactions, and keep the alliance together despite these internal, as well as external, stresses, until a conscious decision is made to terminate the relationship. Relationships provide the flexibility required between the partners to change direction, and make other necessary adjustments to deal with changes in the competitive or external environments.

Beyond the two key roles described, additional staffing will depend on three issues. First, what staffing is required to implement the technology received or to further develop it, as the case may be? This is issue relatively straightforward. The staffing can usually be developed from that established for the technology transfer.

The second and third issues go beyond technology transfer. The second issue focuses on what we are trying to learn from our partner. What is the value we want to get from the relationship beyond the technology per se? At

Table 16-1 Principles for Managing Strategic Technology Alliances

---

1. Establish strong team leadership
2. Foster relationship-building activities
3. Build flexibility into the structure and foster it among personnel
4. Communicate strategic intent, strategies, and programs
5. Foster communication pathways and linkages
6. Build learning into the alliance—staff to acquire competencies
7. Reward acquisition of skills and competencies
8. Assess progress at all levels in the organization
9. Establish criteria for success and periodically assess performance
10. Establish criteria for terminating the alliance and assess periodically

---

what levels of our partner's organization does this value exist? What competencies does our partner have, from which we could learn and benefit?

Having a strategic technology plan in place will help to answer these questions rather than try to deal with them on an ad hoc basis. Once the answers to these questions are developed, structure the alliance to learn! I emphasize this point because Western organizations don't tend to stress the learning aspect of a relationship to the degree that Far East organizations do.

Finally, the third issue deals with the question, what value does the technology recipient bring to the alliance? How does the other party view the value of the alliance? How will we measure success of the alliance? In time, will the technology recipient become more dependent or independent? If the future direction is not desired, how will the technology recipient alter its course?

Managing the Strategic Technology Alliance

The principles for managing the strategic technology alliance, regardless of the level of commitment and complexity, are summarized in Table 16-1. Many of these points have already been covered. Periodic assessment of the performance of the alliance needs to be made at various levels of the respective partners' organizations on a continuing basis. The senior management is usually in the best position to assess how well the alliance is meeting strategic intents, while the operational levels can best assess how well the alliance is operating, and who is learning the most from whom.

As mentioned earlier, Western cultures do not routinely place as high a value on learning within partnerships as Eastern cultures do. Therefore, Western companies need to work harder at formalizing the learning process. As indicated in the second staffing issue discussed previously, staff to acquire new competencies, and be sure the assigned personnel understand what is expected of them. Along with program reviews, the technology recipient needs to conduct its own internal review of progress made to gain the new competencies. The reward system needs to be altered so that the people acquiring these new skills are recognized by the organization. If rewards are not in place, only the few uniquely self-motivated individuals will seek to acquire the desired new skills, while the idea of gaining new organizational competencies will be ignored.

Recognition needs to be given to the idea that strategic alliances are not forever, but usually have a finite life. As part of the periodic review process, the partners need to make separate and collective assessments regarding the state of the alliance. Is the value growing? Is a higher level of commitment warranted to derive greater value? Is the value diminishing? Is it appropriate to decrease the level of commitment? Is it time to terminate the alliance, or allow it to phase out on its own? How does the partner view the alliance today? What is the partner's vision of the future relationship? How does the partner's vision fit with our organization's view. If different, how will this be reconciled?

Criteria for Success

Operational criteria for success, such as quality of relationships, learning experiences, and meeting strategic goals, have been previously covered. But what about the original, higher purpose for seeking technology acquisition and developing the strategic alliance?

Have the gains made changed the technology recipient's strategic position in one or more ways? Possible criteria to measure success are outlined in Table 16-2. Is there tangible evidence of improvement in the organization's competitive position? Has the improvement resulted in increased revenues? What impact have the results had on costs and earnings? Has there been evidence of faster growth than competitors? What impact has there been on market share?

Regarding strategic positioning, does the enterprise now have more or fewer opportunities? Is the organization more or less dependent on this or other partners? Did the organization acquire new skills or competencies as a result of the alliance? From a competitive perspective, has the enterprise gained or lost ground to competitors, in terms of competitive position or competencies?

Table 16-2 Some Possible Criteria for Measuring Success of the Alliance

---

1. Impact on revenue growth
2. Impact on cost reduction/earnings growth
3. Impact on market share
4. More or fewer opportunities
5. More or less dependent on partner
6. Improvement or weakening of competitive position
7. Stronger or weaker competencies

The assessment process itself is important to integrate the learning experience into the enterprise. This experience upgrades the organization's capabilities to manage the next adventure in technology acquisition.

GUIDING PRINCIPLES

Principle #38:   Relationships are the shock absorbers in a strategic alliance.

Principle #39:   Learning should be emphasized and rewarded as a major benefit from a strategic alliance.

Principle #40:   The viability of an alliance needs to be assessed periodically against preestablished criteria for success.

17

Conclusion

With a reassessment of the enterprise's strategic technology plan and positioning as a result of a strategic technology alliance, the process has come full cycle. A well-thought-out technology acquisition, along with an appropriate alliance to facilitate the technology transfer, should improve the organization's strategic position.

A company may see the result in increased revenues, earnings, or market share. If the strategic intent was defensive in nature, the benefit may be that the company retained its position rather than improved it.

On a less tangible level, development of new competencies or strengthening of existing ones should result in greater opportunities to develop new products, processes, or services, as well as, possibly, new markets. If the alliance has been a positive experience, the result may be greater independence or interdependence of the technology recipient.

Finally, one needs to review the strategic technology matrices previously developed. The competitive position should move positively from following to equal or leading in one or more technologies, as well as increase competency.

The management of external technology is an iterative process, just as with internal programs. And so, the process begins again. Strategic technology plans need to be reviewed on a periodic basis, along with competitive position in selected technologies. Selected technologies, themselves, require periodic review and updating. It is normal to add and drop some technologies over a period of several years.

Reassessment of strategic position in selected technologies results in identifying one or more technologies best acquired or developed from external sources, as opposed to internal development. Competitive technical intelligence can help make this assessment and facilitate the external technology search. Subsequently, a determination of the appropriate level of commitment to acquire the technology is made, followed by negotiations, establishment of the strategic alliance, and managing the technology transfer and or development.

Acquisition of external technologies, except in the smallest organizations, is usually not sequential, but rather is characterized by a number of programs proceeding concurrently. As the enterprise continues to gain experience in balancing its internal and external technology programs, an increasing awareness and understanding is developed regarding the importance of interdependent relationships in the increasingly competitive global environment in which organizations must operate today.

GUIDING PRINCIPLE

Principle #41:   For optimum results, competitive technical intelligence and strategic technology planning should be employed as iterative processes, not one-time programs.

Appendix

Summary of Guiding Principles

At the conclusion of each chapter, one or more guiding principles are cited, aimed at capturing the essence of the material covered. Because these principles represent the main ideas and concepts of this book, it is appropriate to pull them together in concluding this work.

The Management of External Technology

Principle #1:   Today, every enterprise needs to avail itself of external technology to leverage internal efforts.

Principle #2:   A cost-effective way to identify new technologies is to first profile your competitors' technology.

An Overview of Competitive Technical Intelligence

Principle #3:   Competitive technical intelligence must be actionable or it becomes "trivial pursuit".

Principle #4:   Senior management must use and promote competitive technical intelligence if it is to become an ongoing activity within the organization.

Focusing the Competitive Intelligence Program

Principle #5:   In designing the competitive technical intelligence search, consider external influences and latent competition, as well as to current competitors within the industry.

Sources of Competitive Technical Intelligence

Principle #6:   To ensure accuracy and completeness, always use more than one source of information.

Patents as a Source of Competitive Intelligence

Principle #7:   A cost-effective way to initiate a patent search for new technology is to first profile competitors' patents.

Principle #8:   Patent trend analyses provide information not only on the size of competitors' efforts but also show whether the trend is increasing, decreasing, or flat.

Principle #9:   Patent trend analyses not only profile competitors' technologies, but also identify the emergence of new technologies.

Principle #10:   Utilizing key words derived from the new technology, one can answer the question, "Who else in the world is working with this technology?"

Scientific Literature and Conferences

Principle #11:   The scientific literature and federally funded research can be excellent sources for early stage technologies.

Principle #12:   Beyond the obvious value of the formal programs, conferences can be excellent sources of competitive technical intelligence through informal networking interactions.

Principle #13:   It is important to have realistic expectations regarding the time and effort requirements, as well as the risks associated with the commercial development of early-stage technology.

Internal Gatekeepers as a Source of Competitive Technical
Intelligence

Principle #14:   Internal gatekeepers are identified, not appointed.

Principle #15:   Internal gatekeepers can be a key source of competitive technical intelligence if they are integrated into the process, and see their role as something more than a paper exercise.

External Gatekeepers as a Source of Competitive Technical
Intelligence

Principle #16:   Key expert panels are a cost-and time-effective
way to access external gatekeepers for technology
forecasting and as a source of competitive technical intelligence.

Lead Users and Strategic Suppliers as Competitive Intelligence
Sources

Principle #17:   Lead users and strategic suppliers offer an excellent opportunity to advance new technology for
competitive advantage through joint programs.

Principle #18:   Lead users need to be identified and differentiated from good customers.

Principle #19:   Find ways to work with lead users, without losing
good customers.

Principle #20:   Strategic suppliers need to be identified on the
basis of their strategic intent and competitive
technology position, in addition to the traditional
criteria of price, quality, and service.

Principle #21:   Strategic suppliers need to be provided with both
strategic and near-term incentives to engage in
exclusive joint programs.

Organizing the Competitive Intelligence Activity

Principle #22:   The easiest way to institutionalize competitive
technical intelligence is to attach it to an existing
planning process.

Principle #23:   Regardless of the organizational configuration, ultimately one person needs to have all the information available so that it can be integrated into a consistent story.

The Role of Competitive Technical Intelligence in the Strategic Planning Process

Principle #24:   Linking competitive technical intelligence to a strategic technology plan is an ideal way to ensure an actionable and proactive process.

Acquisition of External Technology

Principle #25:   In a strategic alliance, seek no higher level of commitment than is necessary to achieve the desired business purpose.

Principle #26:   A strategic alliance is a process, not just a deal.

Principle #27:   The success of a strategic alliance should be measured by how well it achieves its business purpose, not how long it lasts.

Creating the Strategic Alliance: Partner Selection

Principle #28:   Due diligence is critical to understanding the stage of development of the technology, product, process, or service to be acquired.

Principle #29:   Compatibility of the parties' strategic intents is key to a successful strategic alliance.

Principle #30:   Relationships are important; maintaining continuity in interface personnel goes a long way in providing stability to a strategic alliance.

Principle #31:   Strategic intent must not be compromised for the sake of the relationship if the alliance is to succeed.

Negotiating the Strategic Alliance

Principle #32:   Negotiations have already begun when the first contact with the other party is made, regardless of how trivial the contact may seem.

Principle #33: Preparation by the negotiation team is of paramount importance.

Principle #34: Regarding strategic alliances, negotiations need to be structured for a win-win result.

Executing the Technology Transfer

Principle #35: Understanding the elements of technology transfer can help safeguard against the common problem of underestimating the time and effort required to effectively transfer technology.

Principle #36: A key to effective technology transfer is the formation of a technology transfer team by the receiving party.

Principle #37: Acquisition of external technology saves time only if it is aggressively utilized once acquired.

Structuring and Managing Strategic Technology Alliances

Principle #38: Relationships are the shock absorbers in a strategic alliance.

Principle #39: Learning should be emphasized and reward as a major benefit from a strategic alliance.

Principle #40: The viability of an alliance needs to be assessed periodically against preestablished criteria for success.

Conclusion

Principle #41: For optimum results, competitive technical intelligence and strategic technology planning should be employed as iterative processes, not one-time programs.

References

Chapter 1

1. Fusfeld, H. I. Expansion Features of 1950 to 1980, In *Industry's Future: Changing Patterns of Industrial Research*; American Chemical Society: Washington, DC, 1994; pp 87–90.
2. Schultz, J. W. Aluminum Company of America, Southfield, Michigan, Definition of Technology, Unpublished material, 1996.

Chapter 3

1. Porter, M. E. *Competitive Strategy: Techniques for Analyzing Industries and Competitors;* The Free Press: New York, 1980; p 4.

Chapter 4

1. Brenner, M. E. *Competitive Intelligence Review.* **Fall 1996,** *Vol. 7, No. 3,* 20–27.
2. Berkowitz, L., Consultant, Berkeley, N.J., 1993.
3. Paap, J. E. "Technology Management and Competitive Intelligence: Strategies for a Changing World"; Data and Strategies Group, 1995.
4. Ratliff, P. Business Intelligence in the 1990s. Paper presented at the September 1995 meeting of the Chemical Management & Resources Association, pp 8–15.

Chapter 8

1. Coburn, M. M.; Fusfeld, A. R.; Lee, R. K. Key Expert Panels: Providing Solutions to the Most Elusive Problems; Unpublished paper; The Fusfeld Group, Inc., 1996.

Chapter 11

1. Hauser, J. R.; Clausing, D. *Harvard Bus. Rev.* **1988,** *66* (3), 63–73.
2. Goodrich, N. *CHEMTECH,* **1994,** *24*, 15–17.

Chapter 12

1. Porter, M. E. *Competitive Strategy: Techniques for Analyzing Industries and Competitors*; The Free Press: New York, 1980; p 4.

Index